Wild Animals and Leisure

Wild animals form an integral component of the human leisure experience. They are a significant part of the leisure industry and are economically valuable entities. However, as sentient beings, animals also have rights and welfare needs, and, like humans, may also have their own leisure desires and requirements. This collection provides an in-depth analysis of the rights and welfare of humans and wild animals as the two relate to one another within the sphere of leisure studies. It examines a wide array of animals, such as wolves, elephants, dolphins and apes, in a diverse range of leisure settings in international locations, from captive wild animals in zoos, hunting, swimming with dolphins and animals used as educators and for tourist entertainment. This book provides a forum for future considerations of wild animals and leisure and a voice for animal welfarist agendas that seek to improve the conditions under which wild animals interact with and are engaged with by humans.

Neil Carr is at the University of Otago and is the Editor of *Annals of Leisure Research*. His research focuses on understanding behaviour within tourism and leisure experiences, with a particular emphasis on children and families, sex, and animals. He has authored and edited several books, including *Dogs in the Leisure Experience* (2014) and *Domestic Animals and Leisure* (2015).

Janette Young lectures in health policy, politics and promotion at the University of South Australia. Her research interests are centred on the human–animal intersection, salutogenesis or what creates health and wellbeing, social justice and public policy. She has a background as a social worker in ageing, and project and policy work across a diverse range of human interest areas. As a social work student many years ago, Janette learnt that seeking to holistically meet the needs of some people has to encompass caring about the animals people care about.

Routledge Ethics of Tourism

Series edited by Professor David Fennell

This series seeks to engage with key debates surrounding ethical issues in tourism from a range of interdisciplinary perspectives across the social sciences and humanities. Contributions explore ethical debates across socio-cultural, ecological, and economic lines on topics such as: climate, resource consumption, ecotourism and nature-based tourism, sustainability, responsible tourism, the use of animals, politics, international relations, violence, tourism labour, sex tourism, exploitation, displacement, marginalisation, authenticity, slum tourism, indigenous people, communities, rights, justice and equity. This series has a global geographic coverage and offers new theoretical insights in the form of authored and edited collections to reflect the wealth of research being undertaken in this sub-field.

For a full list of titles in this series, please visit www.routledge.com/Routledge-Ethics-of-Tourism-Series/book-series/RET

1 **Animals, Food and Tourism**
Edited by Carol Kline

2 **Tourism Experiences and Animal Consumption: Contested Values, Morality and Ethics**
Edited by Carol Kline

3 **Wild Animals and Leisure: Rights and Wellbeing**
Edited by Neil Carr and Janette Young

Wild Animals and Leisure

Rights and Wellbeing

Edited by
Neil Carr and Janette Young

Routledge
Taylor & Francis Group
LONDON AND NEW YORK

First published 2018
by Routledge
2 Park Square, Milton Park, Abingdon, Oxon OX14 4RN

and by Routledge
711 Third Avenue, New York, NY 10017

Routledge is an imprint of the Taylor & Francis Group, an informa business

British Library Cataloguing in Publication Data
A catalogue record for this book is available from the British Library

Library of Congress Cataloging in Publication Data
A catalog record for this book has been requested

ISBN: 978-1-138-20928-2 (hbk)
ISBN: 978-1-315-45741-3 (ebk)

Typeset in Times New Roman
by Book Now Ltd, London

Contents

Figures and box

Box

Tables

Contributors

Rachel A. Ankeny is an interdisciplinary teacher and scholar whose areas of expertise cross three fields: history/philosophy of science, bioethics and science policy, and food studies. She leads Australian Research Council funded projects exploring ethical food consumption and attitudes to meat production in Australia.

Robert G. Appleby (PhD candidate, Environmental Futures Research Institute, Griffith University and Co-director Wild Spy Pty Ltd) is an ethologist and technology developer with a keen interest in combining these disciplines to further wildlife conservation efforts. He has been studying dingoes on Fraser Island since 2003 and is a member of the Fraser Island Dingo Working Group. He is currently completing a PhD thesis centring on human–dingo interaction dynamics.

William Boyd is the Professor of Geography at Southern Cross University and the Chair of his University's Human Research Ethics Committee. He is a multi- and trans-disciplinary scholar, a geographer and educationalist, with scholarly interests in human–environment interaction, environmental management and education.

Gay A. Bradshaw is founder and director of The Kerulos Center (www.kerulos.org). Her discovery of Elephant post-traumatic stress disorder (PTSD) led to the founding of the new field of trans-species psychology.

Heather J. Bray is a researcher at the University of Adelaide exploring attitudes to food production, in particular the role of science and technology in agriculture (e.g. genetically modified crops), and ideas of ethical food production and consumption, including attitudes to farm animal welfare.

Susanna Curtin is a senior academic at Bournemouth University. Her principle fields of research concern the human dimensions of wildlife watching and the management of wildlife tourism and conservation. She is particularly interested in the ethical treatment of animals in tourism and the effect of captivity on animal wellbeing. Her research on wildlife tourism has taken her around the world to various wildlife settings and has provided material for journal articles and her travel blog 'rewilding journeys'.

Monika Ferguson is a Research Associate at the University of South Australia, currently supported by funding from the Sansom Institute for Health Research. Her PhD research investigated the impacts of human–animal interactions in zoos, for zoo keepers, visitors and various big cat species.

Eleanor Green is currently working in human resource management for the hospitality industry after recently graduating from Bournemouth University. Whilst her work is predominantly people orientated, her passion for wildlife and the ethical treatment of animals in tourism underpin her travel and research interests.

Sebastian Konyn graduated from the University of Adelaide with a Bachelor of Arts (Advanced) in Philosophy and Gender Studies and is currently completing his Honours year. His contribution to this chapter was undertaken as part of a summer research scholarship with the Food Values Research Group.

Carla Litchfield is a Senior Lecturer in Psychology at the University of South Australia and director of the Conservation Psychology and Applied Animal Behaviour Research Group. Her passion is animal behaviour and comparative cognition, conservation psychology (changing human behaviour to be more sustainable), and living in harmony with nature. She is particularly involved with Great Apes, but has also worked with felids, canids, marine mammals and other animals. Her research interests include responsible wildlife tourism and human–animal interactions and she has led wildlife conservation tours to Rwanda, Uganda, Democratic Republic of Congo, Kenya, Tanzania, India and Borneo. She spent a month locked in an enclosure at Adelaide Zoo as part of the Human Zoo in 2007 to experience life in a zoo enclosure first-hand, and she has served as the President of the Board of the Royal Zoological Society of South Australia. She loves spending time with her daughter and their amazing indoor Bengal cats and whippet.

Tom Mordue is the Norman Richardson Professor of Tourism at Newcastle Business School. His main research interests are in tourism development, culture and space. Tom has published a number of times on angling, including articles in: *Annals of Tourism Research, Current Issues in Tourism,* and the *International Journal of Tourism Research.*

Jessica Bell Rizzolo is a Ph.D. Student in Sociology at Michigan State University with specializations in Animal Studies, Environmental Science and Policy, and Conservation Criminology. She holds a Master of Arts in Psychology from Northwestern University. Jessica's research areas include trans-species psychology, the sociopolitical dynamics of conservation initiatives, wildlife tourism, and the illegal wildlife trade. Her publications include 'There Is No Wild: Conservation and Circus Discourse' (*Society & Animals,* 2015), 'Ideology, Subjectivity and Mind in Animal Models and Infant Research' (in *Animals in Human Society: Amazing Creatures who Share our Planet,* 2015), and encyclopedia articles on elephants and the ivory trade (in *Humans and*

Animals: A Geography of Coexistence, 2017). In addition, she is currently the Director of the Asian Elephant Program at the Kerulos Center, where she conducts research, training, and outreach on psychological indicators of elephant trauma and health in wild and captive settings across Asia.

Sheila Scutter has been committed to animal welfare during an academic career focused on health sciences and education teaching and research. Thus Sheila has raised injured or orphaned possums and other native animals and birds, and volunteered with both the RSPCA and Free the Bears Fund. Recent retirement has allowed Sheila to pursue her passion for animal welfare by increased involvement with the RSPCA and FTB fund.

Bradley P. Smith is a Senior Lecturer and researcher in psychology at Central Queensland University. He specialises in the cognition and behaviour of non-human animals, as well as the human–animal relationship. In 2015 he authored 'The dingo debate', a book exploring the story of Australia's most controversial animal, the dingo.

Yvette Wijnandts is a PhD researcher at the University of Adelaide. Using postcolonial theory and critical feminist theory, she aims to unravel how different structures of power intersect, specifically those of culture and religion, in discourses surrounding the food animal industry in a globalizing world.

Samantha Wilkinson is a Lecturer in Human Geography at Manchester Metropolitan University. Prior to this, she worked as a Research Fellow at The University of Nottingham, on a project that aims to broaden understanding of good home care for people with dementia. Before embarking on this post, she undertook a Human Geography PhD at The University of Manchester. She has presented findings from her both her Research Fellow post, and her PhD research into 'Young People, Alcohol and Urban Life', at a number of international and interdisciplinary conferences, and has written several book chapters, along with journal articles for Children's Geographies, Social and Cultural Geography and Geography Compass. Samantha also completed a Human Geography BA and an Environmental Governance MSc at The University of Manchester. Her MSc thesis aimed to move towards non-anthropocentric environmental governance, by giving voice to a captive chameleon.

Sharon Wilson is Programme Leader of Events Management in the Department of Tourism, Hospitality and Events at the University of Sunderland. Her broad research interests are culture and leisure interactions. Currently, she is researching the mobilities of VW Campervan subcultures in relation to time, motion, affect and dwelling. Prior to joining Sunderland, Sharon spent ten years working in strategic commissioning in the cultural sector with institutions such as London's V&A, Tate and Baltic Centre for Contemporary Art Gateshead.

Rachel Yerbury is a PhD candidate at Southern Cross University. As a practising counselling psychologist, she has an interest in the intersection between animals, the natural environment and human psychological wellbeing.

Acknowledgements

A big thank-you is owed to Faye Leerink in her role at Routledge for encouraging us to develop not one, but two edited works on animals and leisure (this book being partner to *Domestic Animals, Humans, and Leisure*, which is also edited by us). An equally big thank-you is, of course, due to all the authors who have worked on the chapters; without them there truly would be no book. The idea for both books and most of the associated chapters stemmed from a series of sessions we put together for the 2015 Australian and New Zealand Association for Leisure Studies (ANZALS) Conference. Consequently, thanks are also due to ANZALS and the conference organisers for providing the opportunity for the seed of an idea to germinate into a book (or two).

For Neil, as ever, special thanks go to his wife Sarah and children Ben, Tat and Gus. All the animals that have been part of his life (and most especially his dogs – Snuffie, Gypsy and most recently Ebony) are owed thanks for keeping him balanced and providing inspiration. It is also perhaps time to thank the National Parks authorities in Queensland, Australia for giving, unwittingly a push to look into issues surrounding animals and leisure. Without their policy of not allowing dogs in National Parks, it is likely Neil never would have wandered down this research path. Thanks are also due to Durrell (now Jersey Zoo) for helping Neil to move beyond dogs to study all animals in the leisure space. Many other zoos and National Park authorities around the world have helped in this way and Neil thanks them all.

For Janette, a thousand thanks go to Neil as a supportive mentor and colleague in my first foray into book editing. Thanks also to life and work colleagues and friends for tolerance, debriefing, conversations and chocolate as needed. So to Greg, Caroline, Richard and many others (including our authors) 'Thank-you +++'. Because of you and our lovely editorial team my first foray into book editing has been hard, but incredibly fulfilling work.

Finally, for Neil, as the Head of the Department of Tourism at the University of Otago there is the recognition that any research undertaken or works published while an HoD can only occur if the staff of the Department are hugely supportive. In this context, the work of Trudi MacLaren and Pip Lennon is hugely appreciated. Their administrative support gives me the time to be able to focus on things like this book. Thanks also go to all the academics in the Department for being so good at their jobs that they make the HoD job into one that need not consume all my time and energy.

Neil and Janette
August 2017

1 Wild animals and leisure

An introduction

Neil Carr and Janette Young

Introduction

This book provides a forum for the discussion of the position of wild animals in leisure studies and leisure experiences. As such, the chapters within it examine the potential benefits and drawbacks of the positioning of wild animals in leisure environments. These benefits and problems relate to humans and the welfare of individual animals and entire species. All of these issues are grounded in the recognition of the sentience[1] of animals and the implications of this for animal rights and welfare concerns.

Subsequent to the recognition of the sentience of animals it must be recognised that any research agenda for wild animals and leisure that speaks only of the position of them in human leisure is arguably guilty of missing the idea that as sentient beings, animals may also have desire, and/or need leisure. Such leisure may coincide with but should be perceived as distinct from human leisure. This suggests that leisure is not only a right of humans but also of non-human animals. Consequently, a central theme discussed within the book is the right to and need for leisure of wild animals.

The chapters within this book have not been collated simply to provide a display of the current state of research within the field. Rather, the book is designed to be a venue for forwarding knowledge and thinking concerning wild animals and leisure, a tool for championing the need for further work in the field, and a voice for animal welfarist agendas that seek to improve the conditions under which wild animals interact with and are engaged by humans. In this way, while the book is clearly an academic piece of work, it is one that follows contemporary calls for such work to have meaning beyond the boundaries of tertiary education institutes. This meaning applies not just to humans, but to animals as well, helping in the process to give a voice to the voiceless.

This is not the first work to look at wild animals within the human leisure experience. Indeed, a variety of such studies have been published in the past. Most have referred to tourism rather than leisure. This may reflect the relative strengths of leisure and tourism studies in tertiary education institutes over the last 10–15 years. During this period tourism courses and research seems to have gone from strength to strength, while many leisure studies programmes have been closed (Collins, 2017; Fletcher *et al.*, 2017; Silk *et al.*, 2017). In the face of these closures, some commentators have talked about the crisis afflicting or the death of leisure studies (Henderson, 2010; Rowe, 2002; Samdahl, 2000). However, the death of leisure studies is proving to be an erroneous thought as we have seen leisure studies as a research field demonstrating rude health in the last couple of years (Fletcher *et al.*, 2017). This may confirm Henderson's point (2010) that the occasional crisis is no bad thing for a field of study to experience and reflect on. A sign of this rude health is the launching of a new leisure studies journal (*International Journal of the Sociology of Leisure*). In addition, publishers are producing leisure studies book series (see, for example, the *Leisure in a Global Era* series published by Palgrave Macmillan), and the vibrancy of conferences such as the one run annually by the Leisure Studies Association in the UK seems to be increasing.

It is not just the increasing health of leisure studies as a research field that suggests the need for this book to be focused on wild animals and leisure. Rather, it is also related to the understanding that while tourism is arguably a part of leisure, leisure is a larger concept. Consequently, a focus on only tourism and tourists misses a wider population (leisured people) and the industries that cater to them. In addition, leisure studies as an academic area of endeavour offers a rich conceptual history that can provide a foundation upon which to develop understandings of the experiences of people at leisure in relation to wildlife, and the leisure experiences, needs and desires of wild animals. This is arguably not surprising when it is recognised that conceptualisations of leisure can be traced back to classical Greece and the work of people such as Plato and Aristotle (Juniu, 2000; Kelly, 1981; Murphy, 1974).

Today, leisure is conceptualised as a complex and multi-faceted entity that encapsulates measurements of time and is often associated with specific activities or types of activities. However, it is at root more concerned with the concept of freedom, encapsulating both *freedom from* external forces and actors, and the *freedom to be*. These freedoms link leisure to the concept of liminality and issues associated with discovery of the self (Carr, 2017).

Previous works on wild animals and tourism have tended to focus on particular animal types, specifically those that are most attractive to people. Small (2012) and Carr (2016), amongst others, have identified such animals as being large, intelligent, cute and cuddly, and exotic and rare. Those animals most favoured by humans have been identified as 'flagship, charismatic, iconic, emblematic, marquee and poster species' (Small, 2011: 232). It may therefore be no surprise to see that the focus of tourism research on wild animals has been on ones like polar bears (Lemelin, 2006; Lemelin *et al.*, 2008), whales (Catlin & Jones, 2010; Davis *et al.*, 1997; Ziegler *et al.*, 2012) and dolphins (Lusseau & Higham, 2004; Orams, 2004). While reptiles and invertebrates have been identified as being unattractive (Cushing & Markwell, 2011; Small, 2012), there have been some studies of them within the tourism experience (Lemelin, 2007). Similarly, while Moss and Esson (2010) have suggested that birds may be unattractive to people, there has still been some work published on them in relation to tourism (Connell, 2009; Steven *et al.*, 2015). However, it is clear that the focus of wildlife tourism work has been on the most attractive animals, leaving many under-studied.

The need for this book is not just grounded on an argument of the dearth of studies of wild animals in relation to the broad field of leisure, or the relative lack of engagement with the conceptual richness encapsulated by leisure studies by those looking at wild animals, in their own right and in relation to humans. Rather, it is also related to the myriad ways in which wild animals are a part of human leisure environments and experiences. The variety of ways in which wild animals are a part of human leisure and the issue of animal sentience that leads us to question the leisure experiences and needs of wild animals are discussed in the following sections of this chapter. However, before we can get to these sections we must first of all identify what we actually mean by 'wild animals'.

What is a 'wild' animal?

Young and Carr (forthcoming, a) have provided an extensive definition of domestic and domesticated animals. This is important, as wild animals have been defined in opposition to their domesticated counterparts. For example, Usher (1986: 4) has defined wildlife as 'a collective noun relating to non-domesticated species of plants, animals or microbes'. However, this type of definition fails to take into consideration the process of domestication. As a result, it sets up a binary state between domesticated and wild animals, whereas Carr (2015) has noted the existence of a continuum between wild and domestic animals. This continuum recognises that domestication is a process: a journey which some species and individual animals have travelled further along than others. Consequently, while we may see differences between animals we can define as domestic or wild that reflect a binary divide, we can also observe similarities which hint at a continuum filled with fuzzy divides between wild and domestic animals. In this way within individual species we may observe wild and domestic animals, as well as those that exist somewhere in between. Take, for example the pig. Most commonly we may define this animal as being domesticated – certainly a component of a human environment, but not domestic in the sense of living in the human home. However, there are also feral pigs – animals whose predecessors were domesticated but who now live a wild life beyond day-to-day contact with humans. Furthermore, wild boars still exist today, reminders of what pigs were like before they were domesticated. At the opposite extreme, we see some breeds of pigs, particularly pot-bellied pigs, being kept as pets. Such animals have arguably passed across the divide between domesticated and domestic to become part of the human family and live in the family home.

The example of the pig raises the question of whether place matters when defining an animal as 'wild'. Does the animal need to live in the wild to be wild itself? Such a question leads automatically into the thorny question of what is a wild space and then to the equally problematic question of whether any such places actually exist in the world today. This chapter is not the place for an elongated analysis of these questions. Rather, it is sufficient to suggest that a 'wild space', meaning one untouched by humans, is, today, like Utopia, potentially very desirable, at least in the sense of us wishing that it existed rather than having to be in it, but ultimately unreal. Instead, it is recognised that there is no corner of the planet untouched by humans in one way or another, even if the extent of that touch varies dramatically. This brings us to the recognition that just as there is a continuum rather than a binary situation between wild and domestic animals, the same is true of wild, domesticated and domestic spaces. The two continuums are arguably interrelated, with the physical space an animal occupies influencing its positioning on the wild–domestic continuum. This allows wild animals to exist in spaces other than the wild. Newsome *et al.* (2005: ix) have utilised this in their definition of wildlife tourism: 'It can take place in a range of settings, from captive, semi-captive, to in the wild'. Yet such a definition is problematic as it suggests that only in the wild are wild animals free, whereas in any other space they are captive, held against their will. It also ignores the increasing number of non-domesticated (wild) animals that exist in the urban centres of the world. These include hawks, foxes and deer amongst many others. Such animals are clearly not domestic and have learned to live alongside humans rather than being domesticated by them.

It is at the fuzzy edges along the continuums that the most problems exist. At what point does an animal cease to be wild and start to be domesticated? This question raises its head within this book when looking at the experiences of animals within zoos and sanctuaries. These are ostensibly wild animals but they are situated within spaces clearly constructed and controlled by humans. This question is potentially most clearly brought to light in Chapter 8, by Scutter and Young, where the focus is on bears rescued from bile farms and other forms of exploitative captivity who cannot be released back into the wild because they would be unable to survive in their 'natural' habitat. Yet, just because such animals may exist at the margins of any fuzzy divide between being wild and domesticated does not make them any less important. These animals have been positioned in this book rather than in its twin, which is focused on domesticated animals (Young & Carr, forthcoming, b), simply because as a species they are most often associated, in the eyes of people, with the wild rather than the domestic sphere.

Overall, the important point is not how we define wild, domesticated or domestic animals but the recognition that all three are related along a continuum. In this way the issues that influence and impact upon animals is much more important than how we as humans categorise them. Yet this does not make such categorisation unimportant. Rather, the categorisation and the processes that underlie it is another human-constructed influence upon the animal that needs to be considered when examining the experiences of animals in the human leisure experience and human constructions and perceptions of animal's leisure needs.

Animal sentience, welfare and rights

In order to determine whether animals have rights and welfare needs we must first ask whether they are objects or sentient beings, as defined earlier in this chapter. This is a contentious question that has been analysed in detail elsewhere

(Carr & Broom, forthcoming; McConnell, 2005). Increasingly, the evidence suggests that animals are not automated objects incapable of feeling or thinking for themselves. Instead, the evidence is increasingly aligning with the notion that animals are sentient beings (Bradshaw, 2011; Goodall, 2007). This is not to say that they are (merely) the same as humans. As noted in Carr (forthcoming), such a view degrades animals, ignoring the differences that exist across the species (human and non-human) divides. While animal sentience clearly exists in its own right, it is also a human construct as well. This is because animals, and their sentience, exist within a human-dominated and controlled world. Animals, and what they are and what they are capable of, are defined and moulded, both literally and figuratively, by humans. As such, it is important to recognise that social constructions of animal sentience, like any other socio-cultural construct, are temporally and spatially specific (Carr, 2014). In other words, how we as humans view animals and recognise their sentience is prone to change over time and is not constant across different place-based cultures.

All of the authors in this book clearly support the notion that animals are sentient beings. From this standpoint stems a recognition that animals have specific welfare needs and rights. These needs and rights encompass individual and species wellbeing in relation to their involvement in human leisure. In addition, they raise questions about the leisure needs of animals. These issues have been discussed in Young and Carr (forthcoming, b) and encompass the notion forwarded by Broom (2010) that rather than focusing on human constructs of animal needs we must place emphasis on our obligations to animals and the empowerment of animals. This means casting aside humancentric constructs of animals' rights and welfare, and actively listening to what animals tell us they actually need. In this way, the book and the individual chapters within it reject the anthropocentric view that gives no value to the animals beyond that which may accrue to humans who own and/or utilise them. In contrast, they may all be identified with a biocentric perspective that 'gives moral standing to all living things' (McLean & Yoder, 2005: 136).

Wild animals and leisure

The importance of this book is partially related to the previously noted dearth of work on wild animals and leisure, and the growing recognition of the sentience, and consequent welfare needs and rights of animals. However, it is also related to the significant role that wild animals play in human leisure and how this leisure impacts upon, directly and indirectly, a vast array of species and individual animals. Carr and Broom (forthcoming) have talked in detail about the ways in which animals are embedded in and influenced by the tourism experience and this is only multiplied when incorporating all leisure experiences rather than just those associated with tourism. Wild animals are influenced due to their attractiveness to humans, which makes them an integral part of leisure experiences. Those animals which we hate or even simply give no conscious thought to are also influenced by the human leisure experience as they may be barred entry to it and the environments in which it takes place, or victimised within human leisure (e.g., the hunting of feral animals or 'pests').

To think only of the position of wild animals in human leisure is to miss the point that as sentient beings there is a need to consider whether animals have a need for leisure. Furthermore, questions need to be asked about whether such leisure needs are focused around, as noted earlier in this chapter, ideas of time, specific activities, and/or notions of self-discovery and enlightenment. If animal leisure needs and desires are found to not entirely match the human ideals of leisure (at least of which many humans do not strive toward either – see Carr (2017) for a more detailed examination of this point) then this should not be interpreted as being somehow lesser than human leisure. Such a view would be anthropocentric, ignoring the point that 'difference' does not equal 'lesser'.

This book, and the chapters within it, is situated within the reality that animals and humans exist in one space (i.e., the planet Earth) and therefore animal leisure and the position of animals within human leisure are intimately intertwined. Such recognition leads to the realisation, highlighted particularly in Chapter 6 (by Wilkinson), that within this nuanced leisure environment there is also the potential to think of the wild animals who are human leisure attractions as being at 'work' while being the focus of the human gaze. This brings to the study of animals the long-running debate about the relationship and divide between leisure and work in human-focused leisure studies.

Structure of the book

Following this introduction, the second chapter, by Ferguson and Litchfield, provides an overview of human–wild animal leisure experiences. This conceptual chapter explores the myriad benefits humans can gain from experiences with wild animals in the leisure environment and also how animals may benefit from exposure to humans at leisure. In addition, the chapter explores the dangers of human–wild animal experiences in the leisure environment for all participants. As such, Chapter 2 provides a strong foundation from which the rest of the contributions to this book can be viewed. The rest of the chapters are each unique but also overlap with one another in a range of ways. Some focus more on the animals while others pay more attention to the humans. Yet all of them in their own ways talk across the human–animal divide, sharing at heart a concern for animal welfare.

As is the case with any edited book the individual chapters are designed to be able to stand alone and therefore the reader can delve into the book searching for specific information relevant to them rather than reading the entire book from front to back. However, the book has been structured in such a way as to take the reader on a journey through the complexities of understanding the rights and welfare of wild animals in human leisure experiences, to exploring conceptualisations of animals as having leisure of their own if the reader wishes to read all the chapters in sequence.

The book is primarily structured in relation to the context in which the animals find themselves situated. This structure clearly highlights the fuzzy boundaries that exist between wild and non-wild animals and their environments. In this way, Chapters 3 to 6 are focused within the zoo environment. Chapters 7 and 8 then focus on wild animals living in captive environments that do not necessarily fit with traditional definitions of zoos but are not necessarily clearly distinct from them either. The rest of the chapters in the book (Chapters 9 to 12) are focused on wild animals, which at first glance exist in unbounded spaces but upon closer inspection raise questions about the freedom of these animals. This invites an examination of what it means to be a wild animal in a similar way to the critiquing of the removal of bars and other overt markers of captivity in zoos without truly freeing the animals housed within them.

Chapter 3 provides an analysis of current knowledge relating to interactive experiences offered by zoos to visitors. In doing so, the chapter examines the potential positive and negative implications of such interactions with wild animals. As a result, the chapter talks both of species wellbeing (via conservation) and individual animal welfare. The authors, Ferguson and Litchfield, point to methodological weaknesses in this literature, calling for more rigorous research to fully understand the complex impacts, at species and individual levels and across both the short and long term, of these wild animal–human interactions. Noting the fuzzy distinction between zoos and other wildlife settings there is arguably no reason why the implications and calls for research stemming from this chapter should be bounded by the fence that marks the physical end of the zoo as an enclosed leisure space for paying human visitors.

Chapter 4, by Curtin and Green, builds on the preceding chapter to focus on the ethics of keeping wild animals captive in zoos. Specifically, the chapter examines the views of people who had visited an animal attraction in the 12 months prior to the study on which the chapter is based being conducted. The findings arguably reflect wider societal positions that see us torn between a desire to be entertained by viewing wild animals and a recognition of their welfare needs that can appear to be at odds with this desire. The chapter also highlights how respondents possess a dose of scepticism regarding claims made about the promise of conservation of species by wildlife tourism attractions. In Chapter 5, Carr drills down further into the desires of people to interact with animals within zoos and the ethics associated with doing so, and preventing visitors from doing so. Specifically, the chapter focuses on touching and interacting with wild animals in zoos by visitors outside of the oversight and control of zoo guides. Such behaviour is, at least in a western context, today defined as inappropriate or deviant behaviour. Again, this chapter speaks to the overlap between species and individual welfare.

The focus switches in Chapter 6 to the individual animal and her leisure. Wilkinson focuses attention on the life of a captive chameleon called Camilla and talks of her leisure and labour life and the interconnections between them. In doing so, the chapter speaks to the complex relations between leisure and non-leisure that have been a feature of the humancentric leisure studies literature. Yet this chapter, like all the others in the book, is never only about the leisure or labour of animals. As Wilkinson notes, the labour of Camilla provides leisure for humans, but does Camilla herself need space away from this 'labour' for her own leisure? This is related to, but different from, the drive by zoos to provide spaces for animals away from the gaze of the human visitor. This development has been focused on animal welfare, but not on their need for escape from 'work' or the opportunity to engage in leisure.

Chapter 7 continues to focus on the animal, switching to elephants that live outside of the zoo enclosure but are nonetheless captive rather than free. Specifically, Rizzolo and Bradshaw focus on the stresses and traumas suffered by Asian elephants often, though not solely, through their involvement in the tourism industry. The authors also hold out hope that through careful consideration and prioritisation of the elephant, the tourism industry can help to educate people about elephants rather than simply exploit them. In doing so, the chapter links to issues raised in Chapter 3.

In Chapter 8, by Scutter and Young, the focus is on rescued bears and how the rescuing of and caring for these bears is a form of leisure for humans. The importance of leisure for the bears is also a central theme of the chapter that questions whether the care and environment being provided for the rescued bears is a form of leisure and whether it is one that ultimately meets the welfare needs of the bears. In this way the chapter shares many similarities with Wilkinson's earlier chapter on Camilla the chameleon.

Yerbury and Boyd, in Chapter 9 focus attention on a critique of humancentric thinking, with specific reference to human–dolphin interactions in the leisure experience. They talk of the need to abandon the anthropocentric position and adopt an ecocentric position instead. Like all of the previous chapters, Yerbury and Boyd do not call for the end of human–wild animal interactions in the leisure space but for a repositioning of the animals in these interactions so that they are not merely objects for the titillation of humans.

Following the aquatic theme of Chapter 9, Chapter 10, by Mordue and Wilson focuses on recreational fishing, one of the most widespread leisure activities of humans, though arguably under-researched in the leisure studies field. The chapter talks of the social construction of angling and anglers and of their relationship with the fish they hunt. This discussion highlights the potential of a social justice lens to ensure the welfare of fish and anglers alike. In this way, the chapter continues a theme common to the entire book, namely how to enable animal welfare and human leisure within the reality of one world where power imbalances are the norm (with the power residing with humans).

Chapter 11, by Appleby and Smith, returns to one of the questions posed by Wilkinson, and Scutter and Young – do animals have an appreciation and need of leisure? Appleby and Smith address this question while looking at killing undertaken by dingoes in Australia, which at first glance may appear unnecessary from a survival-focused perspective. This is a complex question, made all the more difficult to answer by the problem of having to understand observable behaviour across species; a problem many of those contributing to this book have had to contend with.

Bray, Konyn, Wijnandts and Ankeny, in Chapter 12, retain the focus on hunting but focus on humans as hunters rather than animals. The chapter is focused exclusively on Australia, with the authors arguing that as a socio-cultural practice the nature of hunting is place specific and must be understood as such. In this way, it is not possible to merely generalise ideas about hunting from one place to another. Interesting though this point is, the real issue at the heart of the chapter is the notion of ethical hunting: a contentious notion but one that nonetheless is deserving of attention as it raises complex questions about animal welfare that deal not just with the hunted wild animal but also the farmed domesticated animal.

The concluding chapter seeks to draw together the issues that have emerged from the preceding chapters and chart a way forward for understandings of wild animals in human leisure and their welfare and rights in general and specifically relating to a leisure of their own.

Note

1 Sentience is the ability to feel and experience one's surroundings and experience feelings of wellbeing or the converse. Indeed, Carr (forthcoming) has defined sentience as 'encompassing conscious thought, self-awareness, active agency, and the experience of feelings (including pain and pleasure) as more than just survival mechanisms'.

References

Bradshaw, J. (2011). *In Defence of Dogs: Why Dogs Need Our Understanding*. Allen Lane. London.

Broom, D. (2010). Cognitive ability and awareness in domestic animals and decisions about obligations to animals. *Applied Animal Behaviour Science*. 126 (1): 1–11.

Carr, N. (forthcoming). Tourist desires, and animal rights and welfare within tourism: A question of obligations. B. Grimwood, K. Caton, L. Cooke, & D. Fennell (eds). *New Moral Natures in Tourism*. Routledge. Abingdon, Oxon.

Carr, N. (2017). Re-thinking the relation between leisure and freedom. *Annals of Leisure Research*. 20 (2): 137–151.

Carr, N. (2016). An analysis of zoo visitors' favourite and least favourite animals. *Tourism Management Perspectives*. 20: 70–76.

Carr, N. (2015). Introduction: Defining domesticated animals and exploring their uses by and relationships with humans. N. Carr (ed.). *Domestic Animals and Leisure*. Palgrave Macmillan. Basingstoke, pp. 1–13.

Carr, N. (2014). *Dogs in the Leisure Experience*. CABI. Wallingford.

Carr, N. & Broom, D. (forthcoming). *Animal Welfare and Tourism*. CABI. Wallingford.

Catlin, J. & Jones, R. (2010). Whale shark tourism at Ningaloo Marine Park: A longitudinal study of wildlife tourism. *Tourism Management*. 31 (3): 386–394.

Collins, M. (2017). Looking back at leisure: An abridged version of 'the growth of many leisures? Three decades of leisure studies 1982–2011'. *Leisure Studies*. 36 (2) 163–169.

Connell, J. (2009). Birdwatching, twitching and tourism: towards an Australian perspective. *Australian Geographer*. 40 (2): 203–217.

Cushing, N. & Markwell, K, (2011). I can't look: Disgust as a factor in the zoo experience. W. Frost (ed.). *Zoos and Tourism: Conservation, Education, Entertainment?* Channel View Publications. Bristol, pp. 167–178.

Davis, D., Banks, S., Birtles, A., Valentine. P. &. Cuthill, M. (1997). Whale sharks in Ningaloo Marine Park: managing tourism in an Australian marine protected area. *Tourism Management*. 18 (5): 259–271.

Fletcher, T., Carnicelli, S., Lawrence, S. & Snape, R. (2017). Reclaiming the 'L' word: Leisure studies and UK higher education in neoliberal times. *Leisure Studies*. 36 (2): 293–304.

Goodale, J. (2007). Foreword. M. Bekoff. *The Emotional Lives of Animals*. New World Library. Novato, CA, pp. xi-xv.

Henderson, K. (2010). Leisure studies in the 21st century: The sky is falling? *Leisure Sciences*. 32 (4): 391–400.

Kelly, G. (1981). *Leisure in Your Life: An Exploration*. Saunders College Publishing. Philadelphia.

Juniu, S. (2000). Downshifting: Regaining the essence of leisure. *Journal of Leisure Research*. 32 (1): 69–73.

Lemelin, H. (2007). Finding beauty in the dragon: The role of dragonflies in recreation and tourism. *Journal of Ecotourism*. 6 (2): 139–145.

Lemelin, H. (2006). The gawk, the glance, and the gaze: Ocular consumption and polar bear tourism in Churchill, Manitoba, Canada. *Current Issues in Tourism*. 9 (6): 516–534.

Lemelin, H., Fennell, D. & Smale, B. (2008). Polar bear viewers as deep ecotourists: How specialised are they? *Journal of Sustainable Tourism*. 16 (1): 42–62.

Lusseau, D. & Higham, J. (2004). Managing the impacts of dolphin-based tourism through the definition of critical habitats: the case of bottlenose dolphins (Tursiops spp.) in Doubtful Sound, New Zealand. *Tourism Management*. 25 (6): 657–667.

McConnell, P. (2005) *For the Love of a Dog: Understanding Emotion in You and Your Best Friend*. Ballantine Books. New York.

McLean, D. & Yoder, D. (2005). *Issues in Recreation and Leisure: Ethical Decision Making*. Human Kinetics. Champaign, IL.

Moss, A. & Esson, M. (2010). Visitor interest in zoo animals and the implications for collection planning and zoo education programmes. *Zoo Biology*. 29: 715–731.

Murphy, J. (1974). Philosophical dimensions of leisure. J. Murphy (ed.). *Concepts of Leisure: Philosophical Implications*. Prentice-Hall. Englewood Cliffs, NJ, pp. 1–24.

Newsome, D., Dowling, R. & Moore, S. (2005). *Wildlife Tourism*. Channel View Publications. Clevedon.

Orams, M. (2004). Why dolphins may get ulcers: Considering the impacts of cetacean-based tourism in New Zealand. *Tourism in Marine Environments*. 1 (1): 17–28.

Rowe, D. (2002). Producing the crisis: The state of leisure studies. *Annals of Leisure Research*. 5 (1): 1–13.

Samdahl, D. (2000). Reflections on the future of leisure studies. *Journal of Leisure Research*. 32 (1): 125–128.

Silk, M., Caudwell, J. & Gibson, H. (2017). Views on leisure studies: Pasts, presents & future possibilities? *Leisure Studies*. 36 (2): 153–162.

Small, E. (2012). The new Noah's Ark: Beautiful and useful species only. Part 2. The chosen species. *Biodiversity*. 13 (1): 37–53.

Small, E. (2011). The new Noah's Ark: Beautiful and useful species only. Part 1. Biodiversity conservation issues and priorities. *Biodiversity*. 12 (4): 232–247.

Steven, R., Morrison, C. & Castley, J. (2015). Birdwatching and avitourism: A global review of research into its participant markets, distribution and impacts, highlighting future research priorities to inform sustainable avitourism management. *Journal of Sustainable Tourism*. 23 (8–9): 1257–1276.

Usher, M. (1986). Wildlife conservation evaluation: Attributes, criteria and values. M. Usher (ed.). *Wildlife Conservation Evaluation*. Chapman and Hall. London, pp. 3–44.

Young, J. & Carr, N. (forthcoming, a) Introduction: Domestic animals, humans, and leisure – Rights, welfare, and wellbeing. J. Young & N. Carr (eds). *Domestic Animals, Humans, and Leisure: Rights, Welfare, and Wellbeing*. Routledge. Abingdon, Oxon.

Young, J. & Carr, N. (eds) (forthcoming, b). *Domestic Animals, Humans, and Leisure: Rights, Welfare, and Wellbeing*. Routledge. Abingdon, Oxon.

Ziegler, J., Dearden, P. & Rollins, R. (2012). But are tourists satisfied? Importance-performance analysis of the whale shark tourism industry on Isla Holbox, Mexico. *Tourism Management*. 33 (3): 692–701.

2 Human–wild animal leisure experiences

The good, the bad, the ugly

Monika Ferguson and Carla Litchfield

Introduction

People and animals have co-existed throughout history, yet the relationship between the two has varied over time, as well as across human societies. Orams (2002) explains how this relationship can be considered from three perspectives: animals viewed as gods – that is, superior to humans, with people paying homage and/or making sacrifices to animals (as seen in many indigenous cultures); animals viewed as equals – that is, sharing the same rights and needs as humans within their ecosystems (as seen by some ecologists, animal rights activists, and some eastern religions); and animals viewed as subordinates, for humans to utilise for their benefit (this is arguably the dominant thinking in many cultures today, both western and eastern). Taking the last perspective, humans make use of animals in various facets of life, such as for food, hunting and clothing. Humans have also engaged with animals as part of their leisure time, but there has been much less attention paid to these primarily non-consumptive relationships (Orams 2002, Reynolds and Braithwaite 2001). There are a number of factors that might contribute to this. Firstly, this can be attributed to the fact that although people have long interacted with animals, such as breeding and 'taming' wild animals to be kept as domestic pets and companion animals, it was only when European explorers began to collect 'exotic' animal specimens from other lands that opportunities such as zoos and safari hunting began to emerge (Orams 2002). Hunting parks or menageries also existed prior to this in India and Central Asia, with cheetahs considered 'inseparable royal companions' in the courts of Indo-Islamic cultures (Thapar, Thapar and Ansari 2013). More recently, the increasingly urbanised western lifestyle, and consequent disconnect from the natural world, has been argued to lead to an increase in people seeking out opportunities to reconnect with animals (Curtin 2009, Orams 2002). Coupled with an increase in media representations of wildlife, and product development and marketing by tour operators, there are now countless opportunities for people to engage with wild animals as part of their leisure time (Curtin 2009).

This chapter provides an overview of the settings in which people presently interact, engage with, or otherwise encounter, wild animals as part of their leisure time. We propose that such experiences exist along a continuum from wild to captive: from natural parks, to rehabilitation centres, to zoos, and everything in between. Prominent examples of these opportunities will be described. The chapter then considers some of the good, the bad and the ugly aspects of human–wildlife leisure experiences from both human and animal perspectives, drawing on literature from a diverse range of fields, including tourism and leisure studies, animal welfare, ecology, psychology and mental health. Much of the literature exploring these experiences has typically focused on western, educated and affluent people with a high level of disposable income (Curtin 2009), and therefore it should be noted that this chapter is limited in its ability to provide a comprehensive account of wild animal leisure experiences representative of people across the globe.

The world's wild animals

For the purposes of this chapter, wild animals will be used to refer to all extant, non-domesticated species of non-human animals. This includes species of mammals, birds, fish, reptiles, amphibians and invertebrates. According to the Living Planet Report 2016 (WWF 2016), which draws on data from over 14,000 populations of over 3,700 of vertebrate species, there has been a 58% decline in population abundance between 1970 and 2012. Rates of decline vary between habitats: a staggering 81% for freshwater species, compared to 38% for terrestrial species, and 36% for marine species. Across species, habitat loss and destruction is a major contributor to declining populations. Other threats include species over-exploitation, pollution, invasive species and disease, and climate change. For many species, it is a combination of these factors that is causing their decline. Scientists suggest that we are moving into the planet's sixth mass extinction – the Anthropocene, defined by the escalating effects of humans on the global environment over the past three centuries (Crutzen 2002). Paradoxically, despite this continual, human-induced or anthropogenic decline in wild animals, we are also seeing an apparent increase in the desire for people to engage with wild animals as part of their leisure experiences (Curtin and Kragh 2014).

The continuum of wild-animal-related leisure experiences

The extensive opportunities that exist to encounter wild animals during leisure experiences or wildlife tourism (Reynolds and Braithwaite 2001) make it somewhat difficult to conceptualise this phenomenon. Orams (1996) describes a model for categorising tourist–wildlife interactions, which has been adapted for the purpose of this chapter to include not only tourism experiences, but leisure experiences more broadly. This is based on the recognition of the close relationship between leisure and tourism (Carr 2002). As outlined in Figure 2.1, leisure experiences with wild animals exist along a continuum from wild to captive, defined by the level of 'naturalness' of these settings, and the degree of human influence to develop them. These categories are not mutually exclusive, with some leisure activities potentially occurring anywhere along the continuum

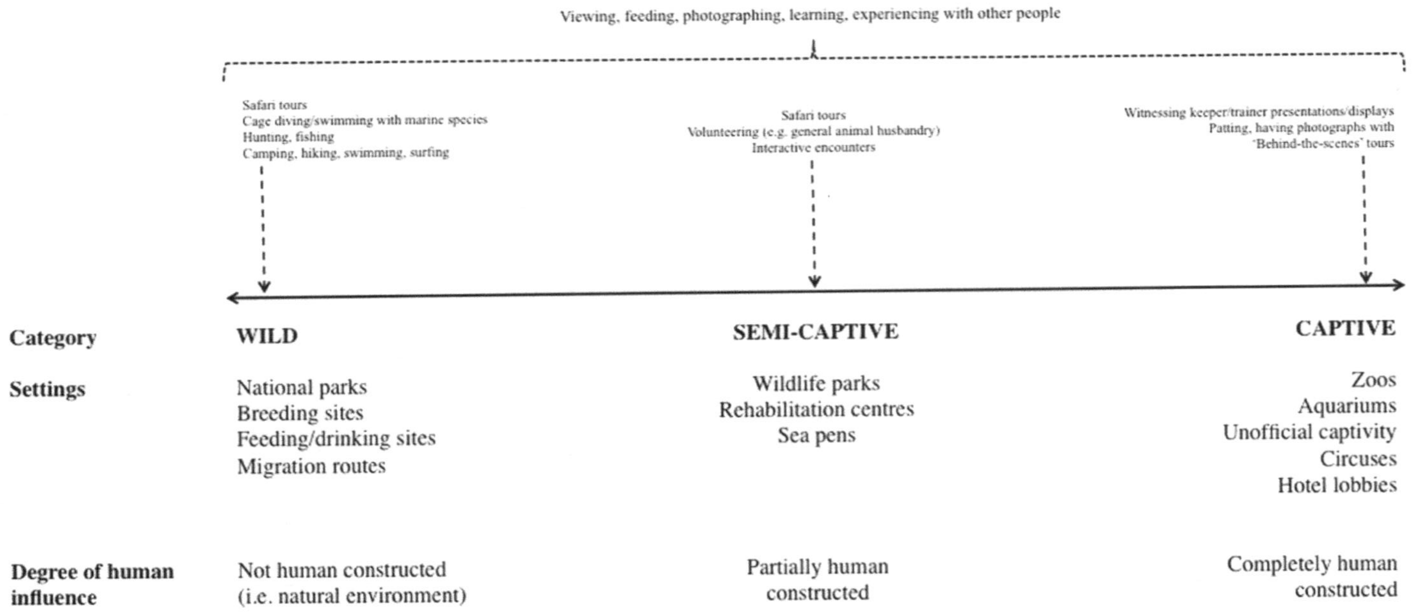

Figure 2.1 The wild–captive human–wildlife leisure continuum, indicating the major categories, settings and degree of human influence (bottom half of figure), and examples of activities undertaken at each (top half of the figure) (adapted from Orams, 1996).

(e.g. wildlife photography and animal feeding), and with the potential for humans to engage in multiple leisure activities within the same category or setting (e.g. viewing, feeding and photographing animals in a zoo). As such, the continuum allows inclusion of leisure activities that do not fit neatly into a single category. For example, cage diving with great white sharks in South Africa or South Australia can be considered as sitting between wild and semi-captive: although it occurs in a natural environment, there has historically been a degree of human influence to attract sharks to the boats (e.g. provisioning; Richards *et al.* 2015).

Just as the definition of leisure is broad and encompasses a range of activities in a range of settings, these wild-animal-related leisure experiences are equally broad – they might occur as part of specific wildlife tourism destinations (e.g. a vacation to Ningaloo Reef, Western Australia, specifically to swim with whale sharks), part of a shorter-term activity (e.g. visiting a zoo), or as a consequences of other leisure experiences, when interacting or engaging with animals might be accidental or secondary, rather than deliberate, such as surfing or hiking. Opportunities for close proximity to animals can also vary within the same setting. For example, Flinders Chase National Park, on Kangaroo Island, South Australia, is a popular destination for sea lion and seal tourism. Within the same national park, visitors can view New Zealand fur seals from approximately 30 m away during a self-guided tour of Admirals Arch, or can have a supervised guided tour close encounter (>10 m) with an Australian sea lion colony on the beach at Seal Bay (Figure 2.2). Further, the cost of wildlife experiences varies – from free, or minimal, in some settings, to thousands of dollars for designated wildlife tourism holidays. Finally, while some of these experiences are consumptive (e.g. fishing and hunting), the majority are non-consumptive. This chapter provides examples of existing opportunities for animal-related leisure experiences along this continuum.

Figure 2.2 Tourists observing Australian sea lions on a guided tour at Seal Bay, Kangaroo Island, South Australia (photo: Carla Litchfield).

Wild experiences

Wild human–animal leisure experiences are characterised by those that occur in natural animal environments. These include national parks, both marine and land-based, animal breeding and feeding sites, and animal migration routes. There are a number of renowned wild marine animal leisure experiences in Australia, such as feeding dolphins at Monkey Mia, Western Australia (Figure 2.3). Marine experiences are popular – for example, the Great Barrier Reef, Queensland, attracts over two million people per year, for activities such as snorkelling and scuba diving, equating to in excess of AU$2 billion, and the use of over 500 commercial vehicles to transport tourists to the reef (Australian Government 2016). These wild-based activities allow people to view, photograph and occupy the same space as wild animals. Other, more passive opportunities include observing and photographing natural events – such as bats emerging from caves for their evening feed, whales migrating to eat or breed, and turtles as they hatch from their eggs and make their way towards the ocean.

Other leisure activities, such as camping, hiking, swimming and surfing, also bring people into contact with animals in natural habitats, even when the primary purpose of the activity is not to interact with animals. This can result in similar experiences, including watching, photographing and interacting with animals. Finally, people engage with animals for leisure in wild spaces for hunting and fishing purposes. A survey of over 90 million US residents found that 13.7 million people hunted and 33.1 million fished in 2011, spending US$144.7 billion on these activities (U.S. Department of the Interior *et al.* 2011).

Figure 2.3 Some wild bottlenose dolphins swim close to shore at Monkey Mia marine reserve in Western Australia, where selected visitors can feed them fish under supervision from a park ranger (photo: Carla Litchfield).

Semi-captive experiences

Wildlife parks, reserves and preserves

Peoples' fascination with wild animals is clearly evidenced by the rise of wildlife tourism and safaris, and the willingness people have to travel to national parks and reserves in the hope of catching a glimpse of a wild animal (Guggisberg 1975) and, more recently, the desire and ability to post photographs of them on social media (Willemen *et al.* 2015). Wildlife tourism is notoriously popular in Africa. Although a somewhat dated statistic (yet still relevant when considering 'modern' interactions), Kenya has been said to attract over one million international tourists annually for wildlife safari and nature-based tourism (Akama and Kieti 2003). In 2011, wildlife safari revenues contributed an estimated US$1.16 billion to the Kenyan economy, representing almost 14% of the Gross Domestic Product (Ogutu *et al.* 2016). Big cats are particularly popular and steal much of visitors' viewing time (Okello, Manka and D'Amour 2008; Figure 2.4).

Similarly, in India, more than one million people are reported to visit tiger reserves every year (Karanth and DeFries 2011). In addition to larger wildlife parks, smaller semi-captive wildlife facilities exist. For example, the Ubud Monkey Forest in Bali, Indonesia, is a designated tourist attraction and conservation area, where people can watch, photograph, feed and touch macaques (although signs warn them of potential injury as a result of aggressive interactions with monkeys and the risk of disease transmission; Fuentes 2012). In another example, large groups of visitors can safely watch and photograph four packs of wolves from behind a double barrier at the Lakota Wolf Preserve, New Jersey, USA (Figure 2.5).

Rehabilitation centres

Wildlife rehabilitation centres primarily care for and rehabilitate wild animals, often from illegal trade. These exist worldwide for a range of species, such as the Bornean Sun Bear Conservation Centre in Malaysia, Elephant Nature Park in

Figure 2.4 Elephants (left, Ngorongoro Crater, Tanzania) and big cats (right, Bandhavgarh National Park, India) are the main attraction for many tourists (photos: Carla Litchfield).

Figure 2.5 Encounters between tourists and monkeys can be risky when interactions are too close at Ubud Monkey Forest (left), while supervised encounters with wolves take place from behind barriers at Lakota Wolf Preserve (right) (photos: Carla Litchfield).

Thailand, Merazonia in South America, and the David Sheldrick Wildlife Trust in Kenya for orphaned elephants (Figure 2.6). At many of these centres, people can visit for a day, like they would a zoo, whereas others target people willing to volunteer their time during lengthier stays (i.e. weeks or months). Like other animal-based tourism experiences, these centres appear to vary significantly in the care they provide to their animals.

Figure 2.6 The David Sheldrick Wildlife Trust is an orphan-elephant rescue and rehabilitation centre which allows visitors to spend an hour watching young elephants feeding and taking a mud bath under supervision of their caregivers (photos: Carla Litchfield).

Figure 2.7 Ngamba Island Chimpanzee Sanctuary in Uganda has a variety of experiences available for visitors to learn more about the plight of chimpanzees in the wild. Tourists seeking a closer encounter with some of the 49 chimpanzees can book an overnight stay and take part in a 'caregiver for a day' experience (photo: Carla Litchfield).

In 2000, the Pan African Sanctuary Alliance (PASA) was created for primate sanctuaries that rescue and care for thousands of surviving victims of the illegal bushmeat and exotic pet trade. Today there are 22 member organisations in 13 African countries, caring for about 2,850 primates (PASA 2015). As a collective, PASA sanctuaries work together to tackle issues related to wildlife protection and law enforcement, animal cruelty, habitat conservation, public education and outreach programmes to empower local communities (www.pasaprimates.org). Many of the PASA member wildlife centres provide opportunities for people to visit, volunteer or conduct non-invasive behavioural research (Figure 2.7). Since these sanctuaries often provide veterinary care, quarantine facilities and care for numerous (sometimes hundreds') individuals, they usually resemble captive facilities (e.g. zoos) but as they are located in habitat countries, some may be able to release individuals back to the 'wild' (PASA 2016).

Sea pens

Sea pens, or sea sanctuaries, are an alternative to full captivity, whereby animals, such as orcas, are given access to a larger 'enclosure' within their natural habitat. Depending on their location, people can visit and view animals in this environment. Some offer opportunities for visitors to enter the pens. For example, in Port Lincoln, South Australia, visitors can swim with tuna in an enclosed ocean pen.

Captive experiences

Zoos and aquariums

Zoos are perhaps one of the most well-documented human–animal leisure venues. Humans have reportedly kept wild animals since the first attempt at domestication (i.e., around 10,000 BC, Kisling 2001; see Alves 2016 for a historical overview of animal domestication). Although zoos have their origins in menageries and similar institutions dating back at least 4,000 years, it was not until the 19th century that the modern zoo was said to have been born (Hosey, Melfi and Pankhurst 2009). This era saw a shift in audiences, with the general public having access to these places, and exotic animal viewing no longer being reserved for royalty and the wealthy elite (Hosey, Melfi and Pankhurst 2009, Kisling 2001). As zoos evolved from ancient menageries to the more naturalistic places many of them are today, the modern zoo appears to have undergone many changes and seen shifts in the way people interact with many species, from passive viewing of animals housed in barren, concrete cages, to more naturalist and immersive exhibits (Baratay and Hardouin-Fugier 2002).

Today's zoos remain highly popular, with over 300 professionally recognised members belonging to the World Association of Zoos and Aquariums

Figure 2.8 Many zoos offer opportunities for close interactions with wildlife, such as a walk through ring-tailed lemur exhibit at Melbourne Zoo in Australia (photos: Carla Litchfield)

(WAZA) (Barongi *et al.* 2015). A study by WAZA estimated that over 700 million people visit zoos and aquaria worldwide on an annual basis (Gusset and Dick 2011), indicating the potential for a large proportion of the public to experience animals in this environment as part of their leisure time. Today, in addition to viewing naturalistic exhibits and reading educational signage at exhibits, visitors can have interactive experiences (Figure 2.8), such as witnessing keeper demonstrations and asking questions, engaging with touch tables and interactive displays, as well as paying to participate in various interactive experiences with a range of species. These can include feeding animals, having photographs taken with them, touching animals, and going on behind the scenes tours. At present, this information is largely found by searching individual zoo websites, with some literature beginning to explore the impact on both animals and people (see Chapter 3 – 'Interactive zoo visitor experiences: A review of human and animal perspectives').

Circuses and 'unofficial' animal-keeping facilities

Other than accredited zoos, there are a range of other opportunities for people to experience captive animals during their leisure time through circuses and 'unofficial' animal-keeping facilities, including hotel lobbies and grounds (Figure 2.9), as well as private ownership within homes and/or residential properties. However, much less is known about these opportunities in the published literature. This may

Figure 2.9 Some hotels provide visitors with animal encounters that are controversial on animal welfare grounds, such as elephant polo in India and a whale shark (top right of image) in a hotel lobby aquarium in Dubai (photos: Carla Litchfield).

reflect the controversial nature of these facilities, with many not being widely publicised, or being deliberately avoided or boycotted on animal welfare grounds. What is known suggests that circuses have allowed people to view exotic animals since well before zoos were available to the general public (Nyhus, Tilson and Hutchins 2010). It is reported that they emerged during the 16th–18th centuries in the form of travelling exhibitions around Europe, and included various wild animal species (Hosey, Melfi and Pankhurst 2009). Historically, tigers and lions featured more than the other large felids in circuses, and, to a lesser extent, so did leopards, along with other popular animals, such as elephants and bears (Guggisberg 1975, Tait 2012, 2016). Circuses were once considered an important part of cultural life in many countries. For example, they were at the height of their popularity in Australia from the mid-1800s to the early 1900s when they were visited by tens of thousands of people each year (Australian Government 2013). However, the emergence of 'new circus' in the late 1970s saw the removal of many animals, a greater focus on animal welfare, and a focus on human physical acts (e.g. acrobatics). Today, opportunities for non-animal circuses are becoming more popular (e.g. Cirque du Soleil).

In addition to circuses, people can engage with wild animals in other forms of captivity, such as though unofficial zoos and private exotic animal ownership. Examples can be found in many South East Asian countries, such as the controversial Tiger Temple in Thailand (Cohen 2013) (Figure 2.10), which has recently been closed after a raid by Thai police (Evans 2016). Like western zoos, these places provide opportunities to view, photograph, touch and/or feed animals. Some also allow for feeding of live prey to carnivores (e.g. Siberian Tiger Park, China; Salem 2015). Again, much of this information is found online,

Figure 2.10 Tiger Temple in Thailand allowed tourists, even very young children, to interact closely with chained-up tigers, typically to take 'selfies' (photo: Adam Litchfield).

rather than in the peer-reviewed literature. For example, until very recently, information could be found on the popular travel website, TripAdvisor, which has now committed to banning ticket sales to attractions that allow physical contact between people and captive, wild or endangered animals (Rushby 2016). Even when a (non-zoo) wildlife tourism activity is considered to have negative animal welfare or conservation impacts, tourists are often unaware of this and even rate these activities highly on TripAdvisor (Moorhouse *et al.* 2015).

The good

Interacting with wild animals can have a number of benefits, particularly for people. Some of these are discussed here.

Connecting with nature

In addition to the previously discussed marked decline in wild animals, we are seeing an increasing disconnect between people and the natural world (for a summary of how relationships with the natural world have evolved over time, see Vining 2003). Since the industrial revolution, there has been "a psychological split between humans and animals by physically removing humans from daily and routine contact with animals" (Curtin 2009, p. 453). As a result, interacting with animals during leisure time provides people with an opportunity to reconnect with nature, whatever the setting. Curtin and Kragh (2014) explain how such experiences provide "a means to re-engage with nature in order to find reprieve from the strains and stresses of modern life" (p. 547). While 'wild' settings offer an opportunity to see animals in their natural habitat, captive settings offer a substitute for wild animal encounters (Mason 2000) and, perhaps more importantly, an opportunity to be in the presence of animals that would not otherwise be seen in the natural environment (e.g. exotic species). In turn, this can have benefits for human wellbeing.

Human wellbeing

Although more prominent in the companion animal literature, there is increasing recognition of the value of wildlife-related leisure experiences for human health and wellbeing. Studies have revealed the positive impacts wildlife experiences can have on mood and emotion. In a qualitative study, Curtin (2009) sought to understand the value of wildlife tourism experiences among 20 tourists on a bird-watching tour to Andalucia, and a whale- and bird-watching tour in Baja, California. She asked how watching wildlife made participants feel, with results including wonderment and awe beyond articulation, experiencing a state of 'flow', and sensual awakening. She also explored the psychological benefits of the same, with key themes including time to stand and stare, voyeurism and contemplation, spiritual fulfilment and feelings of wellbeing. Similarly, studies of zoo visitors are beginning to document the range of emotions experienced during different types of zoo visits – many of these being positive (e.g. Myers, Saunders and Birjulin 2004), but others negative, particularly with regard to how animals are housed (Curtin and Wilkes 2007). In addition to individual experiences, people often engage in wildlife tourism with others, such as their partner, children, other relatives, friends, or as part of an organised tour. This provides potential opportunities for social connectedness, which can have important benefits for human wellbeing (Sandifer, Sutton-Grier and Ward 2015). Further, repeated positive wildlife experiences may have ongoing mental health benefits, similar to those found when people engage with green spaces, including reducing mental fatigue, potential restorative properties, improving one's outlook on life, as well as coping and recovering from stress (Burls 2007).

Contributions to conservation

Engaging with wild animals can contribute to conservation in a number of ways. The most obvious is the potential for revenue-raising, which can be directed to conservation efforts. While some have questioned the ability for captive facilities to contribute directly to animal conservation, for example due to the poor success of reintroduction programmes (Keulartz 2015) or the unlikely claim that captive-bred lions used in 'walk with lions' tourist encounters in Africa will be reintroduced to the wild (Hunter *et al.* 2013), their ability to do so through the funds they provide may be beneficial. A global study by Gusset and Dick (2011) suggests that the world zoo and aquarium community contributes US$350 million to conservation annually.

More indirectly, engaging with animals can result in positive shifts in human conservation behaviours. In particular, greater support for conservation has been attributed to more interactive encounters, along the continuum of experiences. Research suggests that when people engage in interactive animal encounters (such as feeding, touching and close proximity), they report enhanced satisfaction with their experiences, and display increased knowledge, pro-environmental attitudes and long-term behaviour change intentions (Zeppel and Moulin 2008, Visscher, Snider and Vander Stoep 2009). In this way, interactions with flagship species (such as tigers and giant pandas) may contribute to changes in human behaviour for the betterment of a range of species. Tourism operators and conservation organisations should arguably consider expanding the range of focal or flagship species, as tourists in East Africa have been shown to connect emotionally with eight species in addition to the highly promoted 'Big Five',[1] with these tourists indicating intentions to undertake pro-conservations behaviours on behalf of these other species (Skibins, Powell and Hallo 2016). Further, when a scientist or tour leader is present to explain the behaviour of wild animals as it happens, provide information about conservation and scientific discoveries, as well as manage the behaviour and expectations of tourists (such as at the Moorea dolphin tour, French Polynesia, Figure 2.11), it can be an opportunity to promote pro-conservation behaviour without impacting negatively on the animals involved.

Figure 2.11 A dolphin tour in Moorea provides opportunities to watch spinner dolphins from a boat, with a scientist present to ensure that viewing is good for tourists and does not harass dolphins (photos from video: Carla Litchfield).

Benefits for local communities

Like many forms of leisure and tourism experiences, wildlife interactions can provide financial benefits for local communities, across the spectrum of experiences. In a 2006 survey of 20 South African-based wildlife tour operators, Spenceley (2007) found that nearly all reported having a positive impact on local communities, in terms of providing economic benefits through employment and using local services/products, as well as aiding local education, health and conservation initiatives. However, in an earlier study, Sekhar (2003) surveyed the attitudes of local people towards wildlife tourism in Sariska Tiger Reserve, India (180 households across 18 villages): only one quarter of respondents reported benefiting (directly or indirectly) from tourism, with the key benefit being associated with the sale of milk and vegetables to restaurants that catered for tourists; 62% reported wanting to benefit more, particularly through employment and income generated from marketing local products. Consequently, Sekhar (2003) argues that wildlife tourism in South Asia needs to be improved, with a greater focus on the benefits to local people in order for it to be sustainable and successful. Further, communities that become dependent on wildlife tourism are vulnerable when civil unrest or disease epidemics strike (e.g. Ebola), limiting the benefits of these leisure experiences.

Take hunting as another example: if managed properly, and the number of animals hunted is controlled and sustainable (e.g., through permits and quotas), various authors have argued that hunting can significantly raise financial revenue, particularly in Africa (Baldus 2004, Balme *et al.* 2010, Loveridge *et al.* 2010). Mbaiwa (2004) highlights how tourism hunting can generate income for local communities, using Botswana as an example where local people can benefit if hunting concession areas are sub-leased and annual wildlife hunting quotas are sold to safari hunting companies. Termed ‘community-based safari hunting’, this approach is not without its drawbacks but, if local communities are empowered to learn the necessary skills to play key roles in such processes, it is thought to be sustainable and beneficial (Mbaiwa 2004). These benefits extend beyond the financial domain, including environmental benefits (preventing degradation of, and increasing appreciation for, the natural environment by the community in order to preserve it for tourism activities), socio-cultural benefits (preserving cultural and social heritage, and in-turn promoting wellbeing) and building skills (both individual and collective capacity building) (Simpson 2008).

The bad and the ugly

When people interact with animals during their leisure time, the benefits primarily relate to people, whereas the limitations largely concern animals.

Animal welfare

There are concerns for the welfare of animals involved in human leisure experiences, particularly in captive settings. For example, there has been longstanding debate regarding the ethics of using animals in circuses (Tait 2012) and, although scarce, some scientific data indicates the negative impacts of circus life on animal welfare (e.g., see Iossa, Soulsbury, and Harris 2009). Similarly, while unofficial forms of captivity, particularly in South East Asian countries, are renowned for poor animal-keeping practices, these have been subject to less scientific scrutiny. Tigers are a prime example of this form of tourism, with thousands kept captive in Asia (Nyhus, Tilson and Hutchins 2010), and potentially a greater number of tigers held in captivity than there are in the 'wild' (Nyhus, Tilson and Hutchins 2010). Perhaps the most renowned of these interactive organisations was the previously discussed Tiger Temple, in Thailand. The now defunct website of this attraction previously stated that this organisation was established as a rescue centre for poached tigers, which at its peak saw hundreds of visitors per day interacting hands-on with tigers. Photographs taken at the facility, freely available through Google searches (see also Figure 2.10), show chained tigers, exposed to heat and in close proximity to each other, being touched or even sat on by human visitors of all ages (from infants to adults). Other images show use of force and punishment to condition and control the tigers' behaviour. A non-scientific report commissioned by Care for the Wild International (2008) concluded that, despite the initial good intentions of the facility, the animals are "poorly housed and badly maltreated to make them compliant and perform for visitors" (3), that the facility "has become a breeding centre to produce and keep tigers solely for the tourists and therefore the Temple's benefit" (3), and that it participates in illegal international trafficking. Scientific investigations of the welfare of animals involved in unofficial organisations such as these, which are not accredited members of the World Association of Zoos and Aquariums, are lacking. One of the reasons may be that scientists must receive approval from Human and/or Animal Research Ethics Committees to conduct research, whether survey, observational or medical (e.g. taking samples for analysis of cortisol levels) in nature, and part of this approval process requires gaining permission from organisations to conduct research at their facilities. 'Undercover' research would be unlikely to be approved by a Research Ethics Committee and without this approval, scientific journals would not publish the findings.

While existing in zoos means that some animals thrive (e.g. greater longevity, regular provision of food, health care monitoring and lack of predators), others can experience negative impacts to their wellbeing, largely attributed to the marked differences in captive settings compared to natural habitats. For many species, significantly smaller 'territories', unnatural housing compositions, close proximity to other species, provision of food that differs in type, quantity, presentation and nutritional value to that found in natural habitats, expectations to be active during zoo opening hours, and the regular presence of a large number of unfamiliar humans have a negative impact on them. By the very nature of captivity (i.e. to keep animals enclosed in spaces they cannot move from, and under routines determined by people), animals lack the control to behave and engage in activities with the freedom with which they can in their natural habitats. Combined, these factors can have a significant impact on animal wellbeing, with many species displaying stereotypic behaviour – defined as a repetitive behaviour that results from frustration, attempts to cope and/or central nervous system dysfunction (Mason 2006). Examples include pacing, swaying and fur plucking.

Accredited zoos make considerable effort to address these impacts through the provision of environmental enrichment, an approach designed to target physical and psychological wellbeing, by providing variability to an animal's environment by encouraging behavioural opportunities that were not previously available (Swaisgood and Shepherdson 2005). Enrichment can involve a range of approaches, such as manipulation of food type or provision, novel objects or toys, and structural changes to enclosures (for a review of enrichment success, see de Azevedo, Cipreste and Young 2007), and more recently the use of digital interactive technologies (Webber, Carter, Smith and Vetere 2017).

In addition to these individual animal welfare impacts, which are a major concern for animal rights proponents, captivity calls into question the benefits from a utilitarian perspective – that captivity offers a potential vehicle for preserving species for release into natural habitats (Keulartz 2015). In addition to hindering the welfare of individual animals, existing in the captive environment may result in 'captive-adapted strains' of animals that may be less suited to their natural habitats. This may be of concern if these animals are part of breeding programmes for releasing animals to their natural habitats (Kirkwood 2003; Mason *et al.* 2007). For example, captive-bred animals may not recognise and avoid predators (Griffin, Blumstein and Evans 2000), have poor hunting or foraging skills (Rabin 2003), be unable to find mates, or may seek out people as they are associated with food or comfort (Tutin *et al.* 2001). Successful release of social animals may require releasing a group of animals (e.g. 20 chimpanzees released in four sub-groups in Conkouati Reserve in Congo, Tutin *et al.* 2001) or a mixture of wild-caught and captive-bred animals (e.g. African wild dogs, van Dyk and Slotow 2003). Viewing and engaging with animals whose welfare is compromised may impact the visitor experience; it may elicit concern and, in turn, may impact their future decisions to support zoos.

Human injury/fatality

Across the continuum of wild animal leisure settings, there is the potential for injury and fatality of both animals and people. In wild settings, depredation on people is argued to occur when animals are too sick, injured or old to take their usual prey (Loveridge *et al.* 2010), or when they encounter tourists unexpectedly (e.g. Australian killed by an elephant in Borneo in 2011). In captive settings, such incidents have occurred when zoo keepers enter animal cages for cleaning (Schiller *et al.* 2007), when cages are not secured (Hejna 2010), and when zoo visitors disobey instructions (e.g. put their hands into animal enclosures, Murphy, Dempsey and Kneafsey 2007; or enter animal enclosures completely, Bock, Ronnenberger and Betz 2000). Despite numerous reports of human injury and fatality when encountering animals, both on land and in the ocean, it is difficult to know how frequently such instances occur. Some have attempted to understand this. Loe and Roskaft's (2004) review sought to determine rates of attacks on humans by large carnivores by integrating scientific literature with interviews of wildlife managers, researchers and park personnel (n = unspecified). These authors reported that, during the 20th century, estimated rates of attacks for large felids were as follows: 12,599 tiger attacks (in Bangladesh, China, India, Indonesia, Malaysia, Myanmar, Nepal, Russia, Singapore, Thailand and Vietnam); 840 leopard attacks (in India, Nepal, South Africa and Uganda); 552 lion attacks (in India, South Africa, Tanzania, Uganda and Zambia); and 18 puma attacks (in Chile, Canada and USA). While it is unclear how many of these occurred during tourism/leisure experiences, this appears to be a relatively comprehensive attempt to understand how many of these attacks have occurred and under what circumstances. However, when considering these rates spread across a century, for the whole of the human population, the numbers are relatively minimal.

Loss of animal life and population decline

As a consequence of human injury and fatality during such encounters, retaliatory killing of animals often results. This is not limited to deliberate wildlife tourism experiences specifically. Incidental encounters with animals as a result of leisure experiences in animal spaces, such as shark 'attacks' on surfers, can also lead to retaliatory killing. In Australia, this has led to controversial shark-culling efforts (e.g. through shark nets and drumlines). Not only does this have implications for the conservation of the targeted shark species, but also for non-target species 'bycatch' (e.g. other shark species, dolphins and turtles), which threatens whole ecosystems (Australian Marine Conservation Society 2016). This is also true for other species in designated tourism experiences, such as wildlife parks, where many other animals are often killed before the 'offending' one is (Goodrich 2010, Loveridge *et al.* 2010).

The potential for wild animals to harm humans also causes their death in other settings, such as private ownership. A leading example is an incident dubbed the 'Ohio Massacre', which received world-wide media attention. In 2011, a Zanesville, Ohio, resident was alleged to have released his collection of numerous exotic animals (including various species of big cats, bears, wolves and primates) from their cages on his property before taking his own life (Australian Broadcasting Commission [ABC] 2011). According to a media report, this event led to the shooting of the majority of these animals by local law enforcement personnel, reportedly as a means to reduce the danger posed to local people and the apparent lack of facilities in which to house these animals even if they could be caught (ABC 2011). This event sparked public outcry around the world and, regardless of one's view of the way this incident was handled, this is a devastating reminder of what can go wrong when wild animals are maintained as pets (see database of 2,800+ exotic animal incidents in the United States, with 87 people killed, Born Free USA 2017). More recently, there was an outcry when Harambe, a silverback gorilla, was killed after a child fell into his enclosure at Cincinnati Zoo (Litchfield 2016).

In other wild animal leisure experiences, such as hunting and fishing, the direct purpose of the activity is to take animal lives. While it is difficult to distinguish the impact of hunting for leisure from hunting for financial benefit (e.g. poaching), it is clear that this has devastating impacts on most animal populations. For example, the recent *Living Planet Report 2016* (WWF 2016) provides a case study of over-exploitation of African elephants, highlighting an increase in illegal killing of these animals since 2005, with over half of dead elephants (n = 14,606) found to be due to illegal killing. In some regions, the proportion of animals dying due to illegal killing exceeds 70%. Similarly, the public debate concerning wolf conservation and management in the United States has been described as a 'battle' between hunters/farmers and conservationists (Fox and Bekoff 2011). Delisting of wolves in six North American states, after removal of grey wolves from the Federal Endangered Species list, resulted in at least 1,000 wolves being hunted in one season (Morell 2014). For many species, the killing of one animal will have flow-on effects for others. Loveridge *et al.* (2010) explain that there is often a higher demand for adult male lions compared to females (perhaps due to the size and features of male pelts) which can result in infanticide, since when the dominant lion in a pride is removed, succeeding males will often kill any offspring in the pride in order to be able to mate with the lionesses.

Exotic animal ownership and illegal trade

Another under-explored area of concern is the potential for zoos and unofficial animal keeping facilities to contribute to exotic animal ownership, through inappropriate messaging to the public. These leisure experiences allow people to view and engage with exotic animals in ways that would not typically occur in natural habitats. Zoo keepers who work with big cats have expressed concern about the inappropriate messages that interactive practices send to zoo visitors about the dangerousness of these animals and the potential to encourage pet ownership (Szokalski, Litchfield, and Foster 2013). By seeing zoo keepers interact with exotic animals, or having an opportunity to do so themselves, visitors may leave these experiences with misperceptions about these animals. Although it is illegal to keep and trade exotic animals in many countries (such as Australia; Environment Protection and Biodiversity Conservation Act 1999), it is legal to do so in some parts of the USA (Born Free USA 2013). There are obvious dangers involved with owning exotic animals as pets. Case studies in the medical and surgical literature explore, in quite graphic detail, the human injuries and fatalities that have resulted in these environments (e.g. Chapenoire, Camiade and Legros 2001, Cohle, Harlan and Harlan 1990, Lazaraus, Price and Sorensen 2001). There is also a real concern for the welfare of animals involved – both during the transportation process and when housed at their destination (e.g. small enclosures, lack of enrichment, and inappropriate social housing), including unregulated breeding practices (e.g. over-breeding and cross-species breeding) and a surplus of animals which often end up euthanised when they are no longer wanted or able to be cared for (for a detailed report on these issues in the USA, see Animal Protection Institute 2005). Globalisation, and the potential for people to interact with animals as part of leisure experiences while travelling away from their country of origin, means that zoos worldwide have a responsibility to set a good example by discouraging such practices. For example, estimates suggest that approximately one-fifth of visitors (over three million) to Australian zoos are international tourists (Aegis Consulting Australia and Applied Economics 2009), and therefore there is the potential for messages to be spread to visitors who can purchase a wild animal in their home countries.

Outside of zoos and other forms of captivity, the potential for confusion about legal and illegal trade in endangered species and 'mixed messages' impacting on conservation exists (Buckley 2014), when countries such as Namibia serve as both a non-consumptive tourism destination and a destination for trophy hunting, with generous quotas. For example, in 2016 the CITES (Convention on International Trade in Endangered Species of Wild Fauna and Flora) national export quotas for Namibia (CITES 2016) were set as: 150 skins (trophies) and live specimens of cheetahs; 5 adult male hunting trophies of black rhinoceros; 180 trophy tusks from 90 elephants; 250 leopards as trophies; and 25 skins (trophies) of Nile crocodiles. As Buckley (2014) stresses, hunting rare and endangered species potentially sends four inadvertent socio-political messages, which compromise conservation. Firstly, Asian nations find it hypocritical when they are pressured to stop consumption of the same species in traditional medicine. Conservation donors also worry that their funds are subsidising hunting, while corrupt government officials may benefit from issuing permits. Finally, wildlife tourists and tourism operators may avoid these countries or regions altogether. The public outcry that followed the killing of Cecil the lion by an American trophy hunter in Zimbabwe, "the largest reaction in the history of wildlife conservation" (Macdonald *et al.* 2016, p. 27), has shown the emotional and psychological impact that the death of an individual can have on millions of people, whereas conservationists can sometimes focus on species rather than individuals (Nelson *et al.* 2016).

Compassionate conservation approaches do not support lethal methods of managing wildlife populations (Ramp and Bekoff 2015). However, discussion of threats to wildlife and other ethical and welfare issues related to 'conservation trophy hunting' (Aryal, Morley, Cowan and Ji 2016) are beyond the scope of this chapter.

Transmission of diseases between humans and animals

The most serious threat posed by close encounters between humans and wildlife is the potential of cross-species transmission of disease, which may wipe out wild populations that lack natural immunity or expose humans to potentially deadly new viruses (Homsy 1999; Wolfe *et al.* 2005). In 2002, the World Health Organisation announced that more than 30 new diseases had emerged, with international travel and trade assisting the global spread of these emerging infectious diseases (e.g. SARS, HIV and avian influenza H5N1, Schillaci *et al.* 2005; and H1N1 influenza, Karesh 2010). Over 60% of the emerging infectious diseases and pandemics are zoonotic, meaning they have occurred as a result of cross-species transmission of micro-organisms, from animals to humans (Hughes *et al.* 2010), and 75% of these originate from wildlife (see Litchfield *et al.* 2012). Ebola virus disease (EVD) has killed more than 11,000 people in more than 20 outbreaks in Africa, and in some areas in Africa (e.g. Minkebé and Mwagne forests in Gabon) has wiped out 90–98% of chimpanzee and gorilla populations (Leendertz *et al.* 2016).

Canine distemper virus (CDV) is emerging as another potential disease threat to many species in the wild and in captivity, and has the potential to adapt to humans if measles vaccinations cease (Martinez-Gutierrez and Ruiz-Saenz 2016). In 2015, four giant pandas died from this virus at the Shaanxi Rare Wildlife Rescue and Breeding Research Centre in China, and since people can carry CDV without showing symptoms, the State Forestry Administration banned tourists from close contact with all endangered wildlife, including giant pandas at the popular Chengdu Breeding Centre (Hvistendahl 2015). In 2012, Middle East respiratory syndrome (MERS) emerged, with about 30% of the 1,500+ people infected with the MERS coronavirus dying (Wernery, Lau and Woo 2017). It is a zoonotic virus spread from dromedaries (one-humped camels) to people, often with no obvious symptoms in the dromedaries.

Close encounters between tourists and camels occur at markets in the United Arab Emirates, with day-trip itineraries (e.g. to Al Ain) from Dubai sometimes including a visit to a camel market located close to an abattoir, the final destination of many of these camels (a fact rarely disclosed to the tourists). Camels are social ungulates, like horses and giraffes, and nose-to-nose greetings are part of their repertoire of social behaviours (Benhajali *et al.* 2009, Kristal and Noonan 1979). Tourists can emulate this greeting by softly exhaling breath into a camel's nostrils, providing a perfect photo opportunity but also the potential for transmission of viruses, like MERS coronavirus, which may be transmitted by fine droplets from camel noses (Figure 2.12). To the best of our knowledge,

Figure 2.12 Kissing giraffes in Kenya (left) and kissing camels in United Arab Emirates (right) should be discouraged as they provide potential opportunities for transmission of zoonotic diseases (photos: Carla Litchfield).

no public health warnings or precautions are taken before, during or after these close encounters with camels (e.g. no access to hand-washing facilities, no face masks). After stopping over in Dubai, tourists may interact closely with wildlife in other settings and countries. Near Nairobi, tourists can hold a food pellet between their lips, to have it taken by a captive giraffe's large and bristly mouth (Figure 2.12). This hairy kiss provides another potential opportunity for disease transmission, since a coronavirus has been detected in a captive giraffe in Ohio (Hasoksuz *et al.* 2007). After interacting with camels in UAE and giraffes in Kenya, the same tourists may trek gorillas in Rwanda, and chimpanzees in Uganda. Apart from the need to prevent a potential pandemic amongst humans, endangered non-human great apes are particularly susceptible to any viruses we may unknowingly be carrying (Macfie and Williamson 2010).

Conclusion

This chapter has provided an introduction to the world's wildlife and described a continuum for conceptualising the nature of various human–wildlife leisure experiences, from those in natural habitats to those in more artificial, captive environments. The chapter described some of the key strengths and limitations of these experiences, from both human and animal perspectives. As the human population continues to increase and connectivity with the natural world decreases, we are likely to see a continued uptake in all aspects of these human–animal leisure experiences. We need to consider the implications of these holistically, for all people, animals and places involved, if these interactions are to be sustainable and mutually beneficial.

Note

1 The term 'Big Five' was originally used to refer to the most difficult game to hunt in Africa: the lion, leopard, rhinoceros, elephant and buffalo. Today, these remain among the most popular animals to sight during safari tours.

References

ABC (Australian Broadcasting Commission) (2011). Zanesville emotions run high over animal shooting. www.abc.net.au/pm/content/2011/s3344452.htm (20.10.11).

Aegis Consulting Australia, and Applied Economics. (2009). Report on the economic and social contribution of the zoological industry in Australia. www.zooaquarium.org.au/wp-content/uploads/2011/10/12_Socio-Economic-Contribution-of-Zoological-Industry-Report.pdf

Akama, J.S. and Kieti, D.M. (2003). Measuring tourist satisfaction with Kenya's wildlife safari: A case study of Tsavo West National Park. *Tourism Management*. 24 (1): 73–81.

Alves, R.R.N. (2016). Domestication of animals. U.P. Alburqurque and R.R.N. Alves (eds) *Introduction to Ethnobiology*. Springer International Publishing. Switzerland, pp. 221–225.

Animal Protection Institute. (2005). A life sentence: The sad and dangerous realities of exotic animals in private hands in the U.S. http://466a221d5f0081643b32–e5fd6e4345ef06428c08a34c1e533de0.r4.cf1.rackcdn.com/Exotic_Pets_Report.pdf.

Aryal, A., Morley, C.G., Cowan, P. and Ji, W. (2016). Conservation trophy hunting: implications of contrasting approaches in native and introduced-range countries. *Biodiversity*. 17 (4): 179–181.

Australian Government. (2013). Circus in Australia – a way of life for 70 years, 1847–1917. www.australia.gov.au/about-australia/australian-story/early-circus.

Australian Government. (2016). Great Barrier Reef. www.australia.gov.au/about-australia/australian-story/great-barrier-reef.

Australian Marine Conservation Society. (2016). Shark culling. www.marineconservation.org.au/pages/shark-culling.html.

Baldus, R.D. (2004). Lion conservation in Tanzania leads to serious human-lion conflicts: With a case study of a man-eating lion killing 35 people. *Tanzania Wildlife Discussion Paper*. 41: 1–63.

Balme, G.A., Hunter, L.T.B., Goodman, P., Ferguson, H., Craigie, J. and Slotow, R. (2010). An adaptive management approach to trophy hunting of leopards (*Panthera pardus*): A case study from KwaZulu-Natal, South Africa. D.W. Macdonald and A.J. Loveridge (eds). *Biology and Conservation of Wild Felids*. Oxford University Press. Oxford, pp. 341–352.

Baratay, E. and Hardouin-Fugier, E. (2002). *Zoo: A History of Zoological Gardens in the West* (O. Welsh, Trans.). Reaktion Books. London.

Barongi, R., Fisken, F.A., Parker, M. and Gusset, M. (eds) (2015). *Committing to Conservation: The World Zoo and Aquarium Conservation Strategy*. WAZA Executive Office. Gland.

Benhajali H., Richard-Yris, M.-A., Mezzaouia, M., Charfi, F. and Hausberger. M. (2009). Foraging opportunity: a crucial criterion for horse welfare? *Animal*. 3 (9): 1308–1312.

Bock H., Ronnenberger, D. and Betz, P. (2000). Suicide in a lions' den. *International Journal of Legal Medicine*. 114 (1–2): 101–102.

Born Free USA. (2013). Summary of state laws relating to private possession of exotic animals. www.bornfreeusa.org/b4a2_exotic_animals_summary.php.

Born Free USA. (2017). Exotic animal incidents, database. www.bornfreeusa.org/database/exo_incidents.php.

Buckley, R. (2014). Mixed signals from hunting rare wildlife. *Frontiers in Ecology and the Environment*. 12 (6): 321–322.

Burls, A. (2007). People and green spaces: Promoting public health and mental well-being through ecotherapy. *Journal of Public Mental Health*. 6 (3): 24–39.

Care for the Wild International. (2008). *Exploiting the Tiger: Illegal Trade, Animal Cruelty and Tourists at Risk at the Tiger Temple*. Care for the Wild International.

Carr, N. (2002). The tourism-leisure behavioural continuum. *Annals of Tourism Research*. 29 (4): 972–986.

Chapenoire, S., Camiade, B. and Legros, M. (2001). Basic instinct in a feline. *The American Journal of Forensic Medicine and Pathology*. 22 (1): 46–50.

CITES. (2016). CITES National Export Quotas. https://cites.org/eng/resources/quotas/export_quotas.

Cohen, E. (2013). "Buddhist compassion" and "animal abuse" in Thailand's Tiger Temple. *Society and Animals*. 21 (3): 266–283.

Cohle, S.D., Harlan, C.W. and Harlan, G. (1990). Fatal big cat attacks. *The American Journal of Forensic Medicine and Pathology*. 11 (3): 208–212.

Crutzen, P.J. (2002). Geology of mankind. *Nature*. 415 (6867): 23–23.

Curtin, S. (2009). Wildlife tourism: The intangible, psychological benefits of human-wildlife encounters. *Current Issues in Tourism*. 12 (5–6): 451–474.

Curtin, S. and Kragh, G. (2014). Wildlife tourism: Reconnecting people with nature. *Human Dimensions of Wildlife: An International Journal*. 19 (6): 545–554.

Curtin, S. and Wilkes, K. (2007). Swimming with captive dolphins: Current debates and post-experience dissonance. *International Journal of Tourism Research*. 9 (2): 131–146.

de Azevedo, C.S., Cipreste, C.F. and Young, R.J. (2007). Environmental enrichment: A GAP analysis. *Applied Animal Behaviour Science*. 102 (3–4): 329–343.

Evans, S. (2016). Good riddance to Thailand's infamous 'tiger temple'. *The Conversation*. https://theconversation.com/good-riddance-to-thailands-infamous-tiger-temple-60387 (3.6.16).

Fox, C.H. and Bekoff, M. (2011). Integrating values and ethics into wildlife policy and management – lessons from North America. *Animals*. 1 (1): 126–143.

Fuentes, A. (2012). Ethnoprimatology and the anthropology of the human-primate interface. *Annual Review of Anthropology*. 41: 101–117.

Goodrich, J.M. (2010). Human-tiger conflict: A review and call for comprehensive plans. *Integrative Zoology*. 5 (4): 300–312.

Griffin, A.S., Blumstein, D.T. and Evans, C.S. (2000). Training captive-bred or translocated animals to avoid predators. *Conservation Biology*. 14 (5): 1317–1321.

Guggisberg, C.A.W. (1975). *Wild Cats of the World*. David and Charles. London.

Gusset, M. and Dick, G. (2011). The global reach of zoos and aquariums in visitor numbers and conservation expenditures. *Zoo Biology*. 30 (5): 566–569.

Hasoksuz, M. Alekseev, K., Vlasova, A., Zhang, X., Spiro, D., Halpin, R., Wang, S., Ghedin, E. and Saif, L. (2007). Biologic, antigenic, and full-length genomic characterization of a bovine-like coronavirus isolated from a giraffe. *Journal of Virology*. 81 (10): 4981–4990.

Hejna, P. (2010). A fatal leopard attack. *Journal of Forensic Sciences*. 55 (3): 832–834.

Homsy, J. (1999). *Ape Tourism and Human Diseases: How Close Should We Get?* International Gorilla Conservation Programme. Kampala, Uganda.

Hosey, G., Melfi, V. and Pankhurst, S. (2009). *Zoo Animals: Behaviour, Management, and Welfare*. Oxford University Press. Oxford.

Hughes, J.M., Wilson, M.E., Pike, B.L., Saylors, K.E., Fair, J.N., LeBreton, M., Tamoufe, U., Djoko, C.F., Rimoin, A.W. and Wolfe, N.D. (2010). The origin and prevention of pandemics. *Clinical Infectious Diseases*. 50 (12): 1636–1640.

Hunter, L.T., White, P., Henschel, P., Frank, L., Burton, C., Loveridge, A., Balme, G. Breitenmoser, C. and Breitenmoser, U. (2013). Walking with lions: Why there is no role for captive-origin lions *Panthera leo* in species restoration. *Oryx*. 47 (1): 19–24.

Hvistendahl, M. (2015). Captive pandas succumb to killer virus. *Science*. 347 (6223): 700–701.

Iossa, G., Soulsbury, C.D. and Harris, S. (2009). Are wild animals suited to a travelling circus life? *Animal Welfare*. 18 (2): 129–140.

Karanth, K.K. and DeFries, R. (2011). Nature-based tourism in Indian protected areas: New challenges for park management. *Conservation Letters*. 4 (2): 137–149.

Karesh, W.B. (2010). Emerging diseases and conservation: An update on One World – One Health. E. Fearn (ed.). *State of the Wild 2010–2011: A Global Portrait*. Island Press. Washington, DC.

Keulartz, J. (2015). Captivity for conservation? Zoos at a crossroads. *Journal of Agricultural and Environmental Ethics*. 28 (2): 335–351.

Kirkwood, J.K. (2003). Welfare, husbandry and veterinary care of wild animals in captivity: Changes in attitudes, progress in knowledge and techniques. *International Zoo Yearbook* 28 (1): 124–130.

Kisling, V.N. (2001). Ancient collections and menageries. V.N. Kisling (ed.). *Zoo and aquarium history: Ancient animal collections to zoological gardens*. CRC Press. Hoboken.

Kristal, M.B. and Noonan, M. (1979). Perinatal maternal and neonatal behaviour in the captive reticulated giraffe. *South African Journal of Zoology* 14 (2): 103–107.

Myers, O.E., Saunders, C.D. and Birjulin, A.A. (2004). Emotional dimensions of watching zoo animals: An experience sampling study building on insights from psychology. *Curator*. 47 (3): 299–321.

Lazaraus, H.M., Price, R.S. and Sorensen, J. (2001). Dangers of large exotic pets from foreign lands. *The Journal of Trauma, Injury, Infection, and Critical Care*. 51 (5): 1014–1015.

Leendertz, S.A.J., Wich, S.A., Ancrenaz, M., Bergl, R.A., Gonder, M.K., Humle, T. and Leendertz, F.H. (2016). Ebola in great apes – current knowledge, possibilities for vaccination, and implications for conservation and human health. *Mammal Review*. 47 (2): 98–111.

Litchfield, C. (2016). Gorillas in zoos – the unpalatable truth. *The Conversation*. Retrieved from https://theconversation.com/gorillas-in-zoos-the-unpalatable-truth-60249 (1.6.16).

Litchfield, C., Lushington, K., Bigwood, S. and Foster, W. (2012). Living in harmony with wildlife: considering the animal's 'point of view' in planning and design. S. Lehmann and R. Crocker (eds). *Designing for zero waste: Consumption, technologies and the built environment*. Earthscan. London, pp. 181–205.

Loe, J. and Roskaft, E. (2004). Large carnivores and human safety: A review. *Ambio*. 33 (6): 283–288.

Loveridge, A.J., Wang, S.W., Frank, L.G. and Seidensticker, J. (2010). People and wild felids: Conservation of cats and management of conflicts. D.W. Macdonald and A.J. Loveridge *(eds). Biology and conservation of wild felids*. Oxford University Press. Oxford, pp. 161–195.

Macdonald, D.W., Jacobsen, K.S., Burnham, D. Johnson, P.J. and Loveridge, A.J. (2016). Cecil: A moment or a movement? Analysis of media coverage of the death of a lion, Panthera leo. *Animals*. 6 (5): 26–38.

Macfie, E.J. and Williamson, E.A. (2010). *Best practice guidelines for great ape tourism*. IUCN/SSC Primate Specialist Group (PSG). Gland, Switzerland.

Martinez-Gutierrez, M. and Ruiz-Saenz, J. (2016). Diversity of susceptible hosts in canine distemper virus infection: A systematic review and data synthesis. *BMC Veterinary Research*. 12 (1): 78–88.

Mason, G. (2006). Stereotypic behaviour in captive animals: Fundamentals, and implications for welfare and beyond. G. Mason and J. Rushen (eds). *Stereotypic Behaviour in Captive Animals: Fundamentals and Applications for Welfare* (2nd ed.). CAB International. Wallingford, Oxfordshire.

Mason, G., Clubb, R. Latham, N. and Vickery, S. (2007). Why and how should we use environmental enrichment to tackle stereotypic behaviour? *Applied Animal Behaviour Science*. 102 (3): 163–188.

Mason, P. (2000). Zoo tourism: The need for more research. *Journal of Sustainable Tourism*. 8 (4): 333–339.

Mbaiwa, J.E. (2004). The socio-economic benefits and challenges of a community-based safari hunting tourism in the Okavango Delta, Botswana. *The Journal of Tourism Studies*. 15 (2): 37–50.

Moorhouse, T.P., Dahlsjö, C.A., Baker, S.E., D'Cruze, N.C. and Macdonald, D.W. (2015). The customer isn't always right – conservation and animal welfare implications of the increasing demand for wildlife tourism. *PloS One*. 10 (10): e0138939.

Morell, V. (2014). Science behind plan to ease wolf protection is flawed, panel says. *Science*. 343 (6172): 719–719.

Murphy, I.G., Dempsey, M.P. and Kneafsey, B. (2007). Tiger bite in captivity. *European Journal of Plastic Surgery*. 30 (1): 39–40.

Nelson, M.P., Bruskotter, J.T., Vucetich, J.A. and Chapron, G. (2016). Emotions and the ethics of consequence in conservation decisions: Lessons from Cecil the Lion. *Conservation Letters*. 9 (4): 302–306.

Nyhus, P.J., Tilson, R. and Hutchins, M. (2010). Thirteen thousand and counting: How growing captive tiger populations threaten wild tigers. R. Tilson and P.J. Nyhus (ed.). *Tigers of the World: The Science, Politics, and Conservation of Panthera tigris*. Academic Press. London, pp. 223–238.

Ogutu, J.O., Piepho, H.P., Said, M.Y., Ojwang, G.O., Njino, L.W., Kifugo, S.C. and Wargute, P. (2016). Extreme wildlife declines and concurrent increase in livestock numbers in Kenya: what are the causes? *PLoS One*. 11 (9): e0163249.

Okello, M., Manka, S.G. and D'Amour, D.E. (2008). The relative importance of large mammal species for tourism in Amboseli National Park, Kenya. *Tourism Management*. 29 (4): 751–760.

Orams, M.B. (1996). A conceptual model of tourist-wildlife interaction: The case for education as a management strategy. *Australian Geographer*. 27 (1): 39–51.

Orams, M.B. (2002). Feeding wildlife as a tourism attraction: A review of issues and impacts. *Tourism Management*. 23 (3): 281–293.

Pan African Sanctuary Alliance (PASA). (2015). Pan African Sanctuary Alliance 2015 Annual Report. PASA. Portland, Oregon. www.pasaprimates.org/manuals-reports/.

PASA. (2016). *Operations Manual* (2nd Ed.). PASA. Portland, OR. www.pasaprimates.org/manuals-reports/.

Rabin, L.A. (2003). Maintaining behavioural diversity in captivity for conservation: Natural behaviour management. *Animal Welfare*. 12 (1): 85–94.

Ramp, D. and Bekoff, M. (2015). Compassion as a practical evolved ethic for conservation. *BioScience*. 65 (3): 323–327.

Reynolds, P.C. and Braithwaite, D. (2001). Towards a conceptual framework for wildlife tourism. *Tourism Management*. 22 (1): 31–42.

Richards, K., O'Leary, B.C., Roberts, C.M., Ormond, R., Gore, M. and Hawkins. J.P. (2015). Sharks and people: Insight into the global practices of tourism operators and their attitudes to shark behaviour. *Marine Pollution Bulletin*. 91 (1): 200–210.

Rushby, K. (2016). TripAdvisor bans ticket sales to attractions that allow contact with wild animals. *The Guardian*. www.theguardian.com/travel/2016/oct/12/tripadvisor-no-touch-policy-wild-animals-holiday-attractions (12.10.16).

Salem, J. (2015). Inside the gruesome world of the Siberian Tiger Park in China. *News Limited*. www.news.com.au/travel/world-travel/asia/inside-the-gruesome-world-of-the-siberian-tiger-park-in-china/news-story/52738e5a86718afc9407bbc50b668a16 (16.8.15).

Sandifer, P.A., Sutton-Grier, A.E. and Ward, B.P. (2015). Exploring connections among nature, biodiversity, ecosystem services, and human health and well-being: Opportunities to enhance health and biodiversity conservation. *Ecosystem Services*. 12: 1–15.

Schillaci, M. A., Jones-Engel, L., Engel, G.A., Paramastri, Y., Iskandar, E., Wilson, B., Allan, J. S., Kyes, R.C., Watanabe, R. and Grant, R. (2005). Prevalence of enzootic simian viruses among urban performance monkeys in Indonesia. *Tropical Medicine and International Health*. 10 (12): 1305–1314.

Schiller, H.J., Cullinane, D.C., Sawyer, M.D. and Zietlow, S.P. (2007). Captive tiger attack: Case report and review of the literature. *The American Surgeon*. 73 (5): 516–519.

Sekhar, N.U. (2003). Local people's attitudes towards conservation and wildlife tourism around Sariska Tiger Reserve, India. *Journal of Environmental Management*. 69 (4): 339–347.

Simpson, M.C. (2008). Community benefit tourism initiatives – a conceptual oxymoron? *Tourism Management*. 29 (1): 1–18.

Skibins, J., Powell, R. and Hallo, J. (2016). Lucky 13: Conservation implications of broadening "Big 5" flagship species recognition in East Africa. *Journal of Sustainable Tourism*. 24 (7): 1024–1040.

Spenceley, A. (2007). *Responsible Tourism Practices by South African Tour Operators: Survey results from participants at the 2006 Tourism Indaba*. International Centre for Responsible Tourism. South Africa. http://anna.spenceley.co.uk/files/TourOperatorICRTreport_Rev%20May07Distrib.pdf.

Swaisgood, R.R. and Shepherdson, J.D. (2005). Scientific approaches to enrichment and stereotypies in zoo animals: What's been done and where should we go next? *Zoo Biology*. 24 (6): 499–518.

Szokalski, M.F., Litchfield, C.A. Foster, W.K. (2013). What can zookeepers tell us about interacting with big cats in captivity? *Zoo Biology*. 32 (2): 142–151.

Tait, P. (2012). *Wild and Dangerous Performances: Animals, Emotions, Circus*. Palmgrave Macmillan. New York.

Tait, P. (2016). *Fighting Nature: Travelling Menageries, Animal Acts and War Shows*. Sydney University Press. Sydney, Australia.

Thapar, V., Thapar, R. Ansari, Y. (2013). *Exotic Aliens: The Lion and the Cheetah in India*. Aleph. New Delhi.

Tutin, C.E.G., Ancrenaz, M., Paredes, J., Vacher-Vallas, M., Vidal, C., Goossens, B., Bruford, M.W. and Jamart, M.W. (2001). Conservation biology framework for the release of wild-born orphaned chimpanzees into the Conkouati Reserve, Congo. *Conservation Biology*. 15 (5): 1247–1257.

U.S. Department of the Interior, U.S. Fish and Wildlife Service, U.S. Department of Commerce, and U.S. Census Bureau. (2011). National survey of fishing, hunting, and wildlife-associated recreation. www.fws.gov/migratorybirds/pdf/fhw11-nat.pdf.

van Dyk, G. and Slotow, R. (2003). The effects of fences and lions on the ecology of African wild dogs reintroduction to Pilanesberg National Park, South Africa. *African Zoology*. 38 (1): 79–94.

Vining, J. (2003). The connection to other animals and caring for nature. *Human Ecology Review*. 10 (2): 87–99.

Visscher, N.C., Snider, R. and Vander Stoep, G. (2009). Comparative analysis of knowledge gain between interpretive and fact-only presentations at an animal training session: An exploratory study. *Zoo Biology*. 28 (5): 488–495.

Webber, S., Carter, M., Smith, W. and Vetere, F. (2017). Interactive technology and human–animal encounters at the zoo. *International Journal of Human-Computer Studies*. 98: 150–168.

Wernery, U., Lau, S.K.P. and Woo, P.C.Y. (2017). Middle East respiratory syndrome (MERS) coronavirus and dromedaries. *The Veterinary Journal*. 220: 75–79.

Willemen, L., Cottam, A.J., Drakou, E.G. and Burgess, N.D. (2015). Using social media to measure the contribution of Red List species to the nature-based tourism potential of African protected areas. *PloS One*. 10 (6): e0129785.

Wolfe, N. D., Heneine, W. Carr, J.K., Garcia, A.D., Shanmugam,V., Tamoufe, U. Torimiro, J., Prosser, A. T., LeBreton, M., Mpoudi-Ngole, E., McCutchan, F., Birx, D., Folks, T., Burke, D. and Switzer, W.M. (2005). Emergence of unique primate T-lymphotropic viruses among central African bushmeat hunters. *Proceedings of the National Academy of the Sciences of United States of America*. 102 (22): 7994–7999.

WWF. (2016). *Living Planet Report 2016. Risk and Resilience in a New Era*. WWW International. Switzerland.

Zeppel, H. and Muloin, S. (2008). Conservation benefits of interpretation on marine wildlife tours. *Human Dimensions of Wildlife: An International Journal*. 13 (4): 280–294.

3 Interactive zoo visitor experiences

A review of human and animal perspectives

Monika Ferguson and Carla Litchfield

Introduction

Humans have a long history of engaging with zoos as part of their leisure experience.[1] Over time, zoos have evolved from ancient menageries where visitors passively viewed animals, to modern, free-choice learning environments, where opportunities for visitors to have close encounters with animals are now common practice. Searches of zoo websites reveal a range of possible interactive experiences, with a diversity of species. These interactions can include visitors both: (a) watching other people (e.g. zoo keepers) interact with animals, such as during designated zoo keeper talks or education programmes (Figure 3.1); and (b) directly interacting with animals themselves, such as through feeding sessions, patting and having photographs with animals, interacting with animals outside of exhibits and participating in 'behind-the-scenes' visitor tours (Figure 3.2). While the former is usually offered as part of the zoo entry cost, the latter usually requires visitors to pay an additional fee.

From the perspective of zoo visitors, interactive experiences can be important for providing an optimal leisure experience, as well as meeting zoo aims of providing opportunities for learning, and fostering appreciation, care and concern for nature. They also provide zoos with additional financial support, which may be used to contribute to conservation programmes. There is a growing body of literature to indicate that general, non-interactive zoo visits can foster knowledge gain (e.g. Falk *et al.* 2007), positive attitude shifts (e.g. Adelman, Falk and James 2000) and emotional responses (e.g. Myers, Saunders and Birjulin 2004). However, despite positive visitor behaviour change often being touted as the ultimate outcome of a zoo visit, evidence for this is lacking – while there is some support for visitors engaging in increased pro-conservation behaviour after their visit (e.g. Pearson *et al.* 2014), others have found that this is not long-lived (e.g. Adelman *et al.* 2000; Dierking *et al.* 2004). Despite the need for more research to better understand the knowledge–attitude–behaviour link in this informal learning setting, it could be assumed that interactive experiences – which extend the traditional zoo experience – have the potential to have a more pronounced influence on these outcomes.

Figure 3.1 At Taronga Zoo, Australia, visitors can watch zoo keepers interact with a variety of species, such as during public training sessions with elephants (top) and leopard seals (bottom)

Whilst promising for visitors, such experiences raise concerns about the welfare of the animals involved. Over the last 30 years, the literature has paid significant attention to the impact of zoo visitors on animals, particularly seeking to understand changes in animal behaviour as a measure of welfare. Both Hosey (2008) and Davey (2007) have reviewed this literature, which has primarily focused on primates, and conclude that visitors can be considered either a source of stress (evidenced through increased aggression and/or abnormal behaviour, or reduced affiliative behaviour), a source of enrichment (evidenced through increased affiliative and other positive species-typical behaviours) or of relatively neutral impact (i.e. no change in animal behaviour in the presence or absence of visitors). This literature also highlights differences in responses between species, as well as individuals; for example, primates (e.g. monkeys and apes) being more susceptible to negative visitor impacts than felids (e.g. lions and tigers) (Hosey 2008). However, similar to the visitor literature, this evidence has tended to favour investigations of the presence or absence of passive visitors at traditional exhibits, rather than the impact of interactive experiences.

Given the increasing opportunities for interactive experiences, and the lack of consensus about the impacts on the humans and animals involved, it appears timely

Figure 3.2 At Monarto Zoo, Australia, visitors can pay an additional fee to participate in a supervised encounter with a cheetah, where they enter the exhibit, and can pat and have photographs taken with the animals (photo: Carla Litchfield).

to collate the existing evidence. To do this, this chapter provides a narrative review of the peer-reviewed literature to date, focusing first on visitors and then on animals.

Literature search strategy

The literature reviewed here has been collected cumulatively by the authors throughout their research careers, and has been added to by database searching, as well as hand-searching key journals (e.g. *Zoo Biology* and *Visitor Studies*) and reference lists of relevant studies. Studies have been included in this review if they: (a) include a clearly defined and described interactive zoo experience with a 'wild' animal;[2] (b) evaluate the experience from a visitor and/or animal perspective; (c) are peer-reviewed (to ensure only those studies of the highest academic quality and rigour); (d) report on original research (i.e. review articles have been excluded); and (e) are written in English. Given that this a narrative synthesis of the literature, and not a systematic search, the studies included here may not be exhaustive.

Interactive experiences from a visitor perspective

Twelve studies exploring the impact of interactive zoo experiences on visitors have been included in this review. As seen in Table 3.1, half of these studies are evaluations of visitors viewing interactive zoo keeper presentations; two studies examine mixed interactive experiences; and touch pools, interactive exhibits, experiences outside of enclosures, and tours have been explored once each. The species involved in these experiences varies, with aquatic species featuring most frequently (n = 5 studies), followed by felids (n = 3 studies). Half of the studies have been conducted in the USA and four in Australia; two did not specify location. Studies have favoured adult visitors as participants (n = 7 studies) or adults and children (n = 2 studies), with one study of children only (two studies did not mention participant age). Sample sizes range considerably, from 10 to 793 participants, with eight studies conducted at single sites, and the remaining four at multiple sites. Between-groups designs are most common (n = 7 studies), which appears appropriate given the focus on comparing interactive experiences to traditional experiences in most instances. In fitting with this, the majority of studies collected data at single time points (e.g. immediately after a zoo experience; n = 8 studies), with two using pre/post data, and another two studies including follow-up phases. Visitor surveys were used as the primary method in 11 of the studies, either through interviews or self-reporting, with three studies also involving direct observation of visitor behaviour at the zoo. Only one study used a qualitative design (participant interviews). Of those that employed surveys, knowledge, attitudes and behaviour (either intentions or actual behaviour) are common outcome measures (n = 8, 7 and 7 studies, respectively), but there is variation in how these were measured.

Keeper talks

Six studies have explored the impact of witnessing a range of interactive keeper talks on zoo visitor outcomes, with generally positive findings. Three of these studies have demonstrated some impact on knowledge. For example, in Visscher, Snider and Vander Stoep's (2009) study, school children who viewed an interactive rhinoceros training session (n = 20) scored higher on a three-item knowledge quiz than their peers who viewed a fact-only talk (n = 21) or who viewed neither (n = 26). Similar findings were reported for adults who viewed great ape training and research demonstrations in Price *et al.*'s (2015) study. However, lack of details about the type of questions limits interpretation of these results. In a more rigorous investigation by Mellish *et al.* (2016), visitors who viewed a seal show (n = 92) were nine times more likely to report learning something new compared to those who viewed a seal exhibit (n = 88). It should be noted though that relatively high actual knowledge was found, with three-quarters of visitors demonstrating accurate knowledge of marine wildlife entanglement in debris (e.g. fishing nets and plastic), regardless of condition.

Table 3.1 Key methodological details of peer-reviewed studies exploring the impact of interactive experiences from a visitor perspective (least to most interactive)

Interaction type	*Study focus*	*Sample*[a] *and zoo*	*Design*	*Main outcome measure/s*	*Authors*
Keeper presentations	Otter training sessions	389 adult visitors at Zoo Atlanta (USA)	*Between-groups*: talk with training, talk with no training, no talk or training	*Behaviour observations*: Visitor stay time at exhibit *Pre and post self-report questionnaire*: Experience evaluation, perceptions of training, exhibit size and staff qualifications, otters in general and otters as pets	Anderson *et al.* (2003)
	Fur-seal educational shows	180 adult visitors at Melbourne Zoo (Australia)	*Between-groups*: show versus exhibit viewing	*Post-experience self-report questionnaire*: Knowledge, attitudes, conservation behaviour	Mellish *et al.* (2016)
	Great ape research and training demonstration	336 visitors (age = NS) at Lincoln Park Zoo (USA)	*Between-groups*: research demonstration, training demonstration, no demonstration)	*Pre/post self-report questionnaire*: Questions NS	Price *et al.* (2015)
	Interactive bird of prey presentations	175 adult visitors at Healesville Sanctuary (Australia)	*Within-groups*: Post-experience and follow-up	*Post-experience and 5/6-month follow-up interview*: Knowledge of conservation behaviour, behavioural intentions and behaviour since visit	Smith, Broad and Weiler (2008)
	Interactive elephant viewing[b]	355 visitors aged >10 years at Zoo Atlanta (USA)	*Between-groups*: Interactive versus traditional exhibit	*Post-experience questionnaire*: Experiences with elephants *Behaviour*: Signing a petition for/against ivory trade; and returning a solicitation card, either upon exit or post to zoo after visit	Swanagan (2000)
	Black rhinoceros training	67 5th-grade children at Potter Park Zoo (USA)	*Between-groups*: Interactive, fact only, and no training	*Post-experience quiz*: Knowledge	Visscher, Snider and Vander Stoep (2009)

(Continued)

Table 3.1 (Continued)

Interaction type	*Study focus*	*Sample*[a] *and zoo*	*Design*	*Main outcome measure/s*	*Authors*
Interactive exhibits	Keepers in tiger exhibit throughout the day[b]	138 visitor groups (age = NS) at Western Plains Zoo and Dreamworld (Australia)	*Between-groups*: Interactive versus traditional exhibit	*Behaviour observations*: Visitor stay time and engagement with keepers and other information sources *Post-experience interview*: Cognitive, affective and behavioural learning	Broad and Weiler (1998)
Interactions out of exhibits zoo	Keepers walking leopard around zoo	150 adult visitors at Point Defiance Zoo and Aquarium (USA)	*Between-groups*: interactive versus traditional exhibit	*Behaviour observations*: Visitor stay time and efforts to seek information (through reading signage/asking questions) *Post-experience questionnaire*: Cognitive/affective learning, attitudes regarding quality of life and care of the animals	Povey and Rios (2005)
Touch pool	Aquatic invertebrate species touch pool	133 visitors age >13 years at two aquariums (NS)	*Within-groups*: before and after experience	*Pre/post-experience self-report questionnaire*: Knowledge, attitudes and behaviour towards wildlife	Ogle (2016)
Tours	Swim-with-dolphins programme	10 adult visitors at multiple sites worldwide (NS)	*Phenomenological interviews*	*Qualitative interview (2 months – 3 years post experience)*: Questions NS	Curtin and Wilkes (2007)
Mixed experiences	Dolphin shows and interaction programmes	993 adult visitors at six zoos[c] (USA)	*Within-groups*: pre/post/3-month follow-up) *Between groups*: tours, demonstrations, regular dolphin viewing, no dolphin viewing (pre/post/3-month follow-up)	*Self-report questionnaire*: Knowledge, attitudes, behaviour related to dolphins and the marine environment	Miller *et al.* (2013)
	Number of experiences with lions	288 adult visitors at Werribee Open Range Zoo (Australia)	*Cross-sectional*	*Self-report questionnaire*: Cognitive, affective and behavioural outcomes	Weiler and Smith (2009)

Notes

a Sample size based on maximum number of participants (e.g. Povey and Rios (2005) observed 150 visitors, but surveyed 112); NS = not stated.
b Studies included other, non-interactive elements which are not relevant for inclusion here.
c Zoos in Miller *et al.*'s (2013) study = Minnesota Zoo, Brookfield Zoo, Indianapolis Zoo, Texas State Aquarium, Disney's The Seas, Dolphin Connection

Similarly, three studies have demonstrated impacts of keeper talks on visitor attitudes. Anderson *et al.* (2003) found that those visitors who witnessed either otter training ($n = 94$) or otter training with a keeper talk ($n = 104$) reported a more positive experience, and had more positive perceptions of training, exhibit size and staff, compared to those who viewed a keeper talk without training ($n = 97$), or viewed the exhibit passively ($n = 94$). Price *et al.* (2015) and Mellish *et al.*'s (2016) aforementioned studies support this.

Again, three studies explored visitor behaviour change associated with this type experience. Smith, Broad and Weiler (2008) investigated intentions among visitors ($n = 175$) to a bird presentation, finding that 54% of participants reported they would either start an action or increase their commitment to action toward bird conservation. In the six-month follow-up with 38 participants, three reported starting an action and 16 had increased their commitment to an action since their visit. Mellish *et al.*'s (2016) study also explored behavioural intentions, finding that although there was no difference in willingness to change diet through eating sustainable seafood, visitors who viewed the seal show demonstrated a trend towards being more willing to change other behaviours (such as donating money to support marine conservation). Further, those in the show condition reported that their visit would impact their future behaviour, and reported greater confidence in their ability to impact seal conservation through behaviour change.

One study explored actual conservation behaviour. Swanagan (2000) asked visitors ($n = 355$) to an interactive elephant demonstration to self-report their past and current experiences with elephants and gave them solicitation cards, which, if returned, would indicate their willingness to support elephant conservation. Overall, 18.3% of participants returned these, with a higher return rate for those who had had more interactive experiences. Finally, one study by Anderson *et al.* (2003) explored behaviour at exhibits, finding that visitors who engaged with training with a talk or training alone stayed longer than those who viewed a regular talk or viewed the exhibit without a talk (stay times = 492 seconds, 289 seconds, 108 seconds and 133 seconds, respectively).

Viewing interactive exhibits

Interactive exhibits – whereby keepers spend a large portion of zoo opening hours in an exhibit and interacting with animals – provide another opportunity for interactive visitor viewing. Broad and Weiler (1998) found that, compared to visitors who viewed a traditional tiger exhibit ($n = 74$ visitor groups), those who viewed an interactive display (note it also included other elements, such as keeper talks, free brochures, signage and touch tables; $n = 64$ visitor groups) recalled a greater quality and quantity of cognitive (facts about tigers), affective (feelings about tigers) and behavioural (conservation) information immediately after their visit. Although not tested for significance, observations of behaviour revealed that fewer visitors at the interactive exhibit read signage compared to those at the traditional exhibit (15% and 50%, respectively). However, more of the former interacted with keepers to ask questions (22% and 14%, respectively), and spent longer at the exhibit (16 minutes and 6 minutes, respectively).

Interactive experience outside of exhibits

Povey and Rios (2005) explored the impact of an interactive programme, whereby keepers walk clouded leopards around the zoo, offering visitors a chance to encounter them spontaneously and ask questions. Surveys found that interactive visitors ($n = 55$) reported more factual information (i.e. something they had learnt), while the exhibit visitors ($n = 57$) reported more observational information (i.e. something they had seen). Furthermore, interactive visitors reported stronger beliefs about the quality of life and care that the animals received. Observations with a second set of visitors revealed that 45% of those who engaged with this experience ($n = 75$) displayed information-seeking behaviours (namely, asking questions), compared to only 25% of visitors who viewed the exhibit ($n = 75$), whose sole information-seeking behaviour was reading exhibit signage. The former also spent significantly more time observing the animal (185 seconds) compared to the latter (55 seconds).

Touch pools

Through a brief (5-item), pre/post survey, Ogle (2016) found that interacting with touch pools contributes to increased perceived knowledge among visitors ($n = 133$), and that this is associated with having prior experience with zoo/aquarium animals. Further, visitors reported a greater likelihood of taking action to support conservation of aquatic wildlife (i.e. behavioural intentions) after the experience. However, the experience had no impact on how much visitors valued wildlife, or on how strongly they felt about the importance of protecting wildlife (i.e. attitudes). Whilst some of these findings are promising for understanding the value of touch pool experiences, generalisability is limited given the lack of details about the facilities or demographic details about participants.

Interactive tours

A phenomenological study by Curtin and Wilkes (2007) explored the impact of participating in 'swim-with-dolphin' tours worldwide. Through a series of retrospective interviews, participants ($n = 10$) indicated that although such experiences provide an opportunity to experience the grace, size and power of captive dolphins, the experiences can be too staged, short and expensive. Some participants also expressed concerns, particularly regarding the size of dolphin exhibits, dolphins existing in captivity, and the 'tricks' performed. Although not intending to explore attitudes specifically, this is an interesting finding as it highlights the potential for interactive experiences to promote or reinforce attitudes that are critical or unsupportive of zoos.

Mixed experiences

Two studies have compared a range of interactive experiences. Weiler and Smith (2009) found that the higher the number of interpretive experiences that visitors (n = 288) had with lions (including viewing the exhibit, listening to a keeper talk, interacting with a volunteer guide, watching an actor thematic interpretation, and attending a behind the scenes tour to visit the off-exhibit facilities and feed the animals), the higher their reported cognitive (e.g., thought provocation), affective (e.g., attitudes towards nature conservation) and behavioural (e.g., desire to participate in additional interpretive experiences) scores were. Similarly, Miller *et al.*'s (2013) multi-institutional study found that both viewing interactive dolphin shows (n = 462) and participating in interactive tours (n = 331) resulted in short-term knowledge gain, positive attitude change, and increased behaviour change intentions. However, while improved knowledge was the only variable to be maintained by the time of a three-month follow-up for show visitors, all increases were maintained for tour visitors. Further, in both conditions, visitors reported engaging in more conservation behaviours at the three-month follow-up compared to the three-months prior to their visit. The study also highlights the cumulative impact of experiences, with findings showing that the number of past dolphin shows witnessed predicted conservation behaviour. Whilst promising, these findings relate to self-reported, rather than actual, behaviour change.

Combining the visitor literature

Despite variation in the types of interactive experiences reviewed here, this literature offers some indications from survey data that interactive experiences can contribute to favourable shifts in visitors' knowledge, attitudes and behavioural intentions for some visitors at some zoos, at least in the short term, and can increase stay time at exhibits. This extends what is known about the impact of visiting zoos in general (as discussed at the start of this chapter). However, changes in actual behaviour remain unknown. Further, qualitative data from one study (Curtin and Wilkes 2007) suggest that, upon reflection, visitors may view these experiences negatively.

Interactive experiences from an animal perspective

Ten studies exploring the impact of interactive visitor experiences on zoo animals have been included in this review. As outlined in Table 3.2, these have primarily explored visitor tours (n = 5 studies), followed by feeding (n = 2 studies); free-range exhibits, keeper-for-a-day programmes and zoo education programme animals (such as those included in keeper talks or interactive presentations) have each been investigated once. Species examined also vary, with dolphins being most popular (n = 4 studies), as do sample sizes (range, n = 3–59).

Table 3.2 Key methodological details of peer-reviewed studies exploring the impact of interactive experiences from an animal perspective (least to most interactive)

Interaction type	*Study focus*	*Sample and zoo*	*Design*	*Main outcome measure/s*	*Authors*
Programme animals	Programme animal behaviour and physiology	Study 1: 59 armadillos at 17 zoos[a] (USA) Study 2: 10 armadillos,12 hedgehogs, 6 red-tailed hawks at 11 zoos[a] (USA)	Study 1: *Between-groups*: education animals, exhibit animals, and off-exhibit animals Study 2: *Within-groups*: handling for education, no handling, post handling for education	Study 1 and 2: *Behaviour observations*: Undesirable behaviour, feeding/drinking, object examination, self-directed behaviour, locomoting, other active, resting, not visible, not visible presumed resting. *Glucocorticoid metabolites*: Fecal sampling	Baird *et al.* (2016)
Free-range exhibits	Relationship between visitor numbers and animal behaviour and stress physiology	15 kangaroos at Melbourne Zoo and Healesville Sanctuary (Australia)	*Correlational*	*Behaviour observations*: Activity (visitor-directed vigilance, general vigilance, foraging, locomotion, resting, grooming self, grooming others, aggression) and distance from visitor pathway *Glucocorticoid metabolites*: Fecal sampling	Sherwen *et al.* (2015)
Animal feeding	Impact of visitor feeding on lemur behaviour	4 lemurs at Newquay Zoo (UK)	*Within-groups*: during visitor feed, post-visitor feed, during keeper feed, post-keeper feed	*Behaviour observations*: Feeding, social positive, aggression, solitary, vigilance, interacting with keepers, interacting with visitors	Jones *et al.* (2016)

	Impact of visitor feeding on giraffe behaviour	20 giraffes at nine zoos[b] (USA)	*Between groups*: all day feeding, some day feeding, no feeding programme	*Behaviour observations*: Idle, locomoting, eating, ruminating, drinking, socializing, grooming, object-manipulating, time out. Stereotypic behaviour: Object-licking, tongue-rolling, aberrant oral behaviour, pacing	Orban, Siegford, and Snider (2016)
Keeper for a day	Whether animals can distinguish between familiar and unfamiliar humans	1 elephant, 3 giraffes, 3 tapirs, 2 meerkats at Paignton Zoo Environmental Park (UK)	*Within-groups*: familiar keepers, unfamiliar keepers – visitors, no keepers	*Behaviour observations*: States (foraging, locomoting, resting, socializing, autogrooming, out of sight); events (interaction with keeper, other); proximities to keepers	Martin and Melfi (2016)
Visitor tours	Whether dolphins can distinguish between different groups of humans	5 dolphins at Dolphins Plus (USA)	*Within-groups*: adult visitors, children <12 years visitors, children <12 years with disabilities	*Behaviour observations*: Distance to other dolphins and humans; swim speed; contact frequency with visitors; contact distance to visitors	Brensing and Linke (2003)
	Impact of visitor tours on animal well-being	3 dolphins at Marineland (New Zealand)	*Within-groups*: before, during, after tour sessions	*Behaviour observations*: Use of refuge area; frequencies and location of surfacing; behaviour change (other, touch, aggressive, submissive, abrupt, play)	Kyngdon, Minot, and Stafford (2003)

(Continued)

Table 3.2 (Continued)

Interaction type	*Study focus*	*Sample and zoo*	*Design*	*Main outcome measure/s*	*Authors*
	Impact of visitor tours on animals during tours with and without keeper control	22 dolphins at Dolphin Quest, Dolphin Plus, Theater of the Sea, and Dolphin Research Centre (USA)	*Within-groups*: presence or absence of trainer/keeper	*Behaviour observations*: Social interactions with visitors or dolphins (aggressive, submissive, sexual, abrupt, neutral/affiliative); behavioural states (social vs non-social activities; activities under direct trainer control v not under direct trainer control); proximity to other dolphins (within 1m); details (age/sex) of visitors within 1m	Samuels and Spradlin (1995)
	Change in animal behaviour associated with 'behind-the-scenes' tours	3 cheetah, 2 lions, 1 tiger at Monarto Zoo and Adelaide Zoo (Australia)	*Within-groups*: before, during, after tours; tour and non-tour days	*Behaviour observations*: Inactive, active species-typical, feeding, conspecific interaction, human interaction, pacing, aggression, out-of-sight, other; proximity to humans	Szokalski, Foster, and Litchfield (2013)
	Short- (daily) and long-term (monthly) impact of visitor tours on animal behaviour	5 dolphins at Marine Life Oceanarium (USA)	*Within-groups*: before and after tours	*Behaviour observations*: Social behaviour (solitary and dolphin-dolphin); behavioural events (play, swimming, motor movements, orienting, sexual, aggressive, resting, miscellaneous)	Trone, Kuczaj, and Solangi (2005)

Notes

a Zoos in Baird et al.'s (2016) study = Akron Zoo, Bergan County Zoo, Busch Gardens Tampa, Cincinnati Zoo and Botanical Garden, Cleveland Metroparks Zoo, Columbus Zoo and Aquarium, Denver Zoo, Florida Aquarium, Fresno Chaffee Zoo, The Good Zoo at Oglebay, Happy Hollow Park and Zoo, Lincoln Park Zoo, Tampa's Lowry Park Zoo, Memphis Zoo, Oregon Zoo, Peoria Zoo, Point Defiance Zoo and Aquarium, San Antonio Zoo, Seneca Park Zoo, Wildlife Safari, Zoo Atlanta

b Zoos in Orban et al.'s (2016) study = Binder Park Zoo, Brookfield Zoo, Cincinnati Zoo and Botanical Garden, Cleveland MetroParks Zoo, Detroit Zoo, Fort Wayne's Children Zoo, Indianapolis Zoo, Lincoln Park Zoo, Pittsburg Zoo.

The majority of these studies were conducted in the USA ($n = 5$), with two in the UK and Australia, and one in New Zealand. Half of the studies were conducted at single zoos, with the other half taking place at multiple zoos, with larger sample sizes for the latter. The majority of studies ($n = 8$) employed within-groups, repeated-measures designs, with one study using a between-groups design, and one mixed. Consistent with the general visitor impact literature (Davey 2007), all studies employed animal behaviour as the primary outcome measure, with a further two complementing this with physiological measures (specifically, faecal glucocorticoid metabolite (FGM) concentration).[3] Various behaviours have been explored, which is not surprising given the range of species involved, although most studies included species-typical behaviours (e.g. locomotion, resting, feeding), interactive behaviours (e.g. proximity to humans) and indicators of stress (e.g. stereotypic behaviour and aggression).

Zoo education programme animals

Zoos appear to use a diversity of species as education programme animals, with activities including keeper talks, frequent handling by keepers, and interactive demonstrations either at exhibits or around the zoo. In a two-part study, Baird *et al.* (2016) have provided the first examination of this form of interactive experience. In the first, no differences in undesirable behaviours (e.g. pacing, rocking, plucking hair/feathers or picking skin) or FGM concentrations were observed between armadillos (n = varied, with a total of 59) involved in education programmes compared to those not involved in programmes (housed both on- and off-exhibit). Although some differences were found in other behaviours – such as less time resting and more locomotion for education animals – differences were attributed to a range of other factors (including amount of handling, and enclosure size and depth of enclosure substrate), rather than involvement in interactive experiences. Similarly, in the second study, no effect of condition was found on armadillo ($n = 10$), hedgehog ($n = 12$) and hawk ($n = 6$) undesirable behaviour or FGM concentration when handled, not handled or after handling for education, although amount of handling was positively correlated with FGM in all species.

Free-range exhibits

Sherwen *et al.* (2015) explored the welfare of two species of kangaroos ($n = 15$) housed in free-range exhibits, whereby visitors walk through the exhibit without a physical barrier between themselves and the animal(s). For both species, a positive correlation was found between visitor numbers and visitor-directed vigilance, but there was no change in other behaviours (general vigilance, foraging, locomotion and resting), distance from visitor pathway, or FGM concentration. Although causation is difficult to determine, given the correlational nature of this study, these findings support the authors' conclusion that visitor numbers have no negative impact on these animals based on the measures used.

Animal feeding

Two studies have explored the impact of animal feeding programmes. Jones *et al.* (2016) observed lemurs ($n = 4$) to display less aggression (although this was almost minimal) and more time interacting with keepers during visitor feeds compared to during keeper feeds, but no changes after feeding between conditions (feeding included both hand and scatter feeding in both). In a larger, multi-zoo study of giraffes ($n = 20$), Orban, Siegford and Snider (2016) found a positive association between time spent participating in a feeding programme (whereby visitors feed the animals from a platform next to the exhibit) and increased idleness and decreased rumination, but no impact on stereotypic behaviours. Further, when combining time spent in routine feeding with interaction time, animals who spent more time feeding spent less time engaging in oral stereotypic behaviour, suggesting a potential enriching impact of feeding programmes for this species.

Keeper for a day

Martin and Melfi (2016) explored a keeper-for-a-day programme on a range of species (elephant $n = 1$, giraffe $n = 3$, tapir $n = 2$ and meerkat $n = 2$), by comparing animal responses to familiar keepers and unfamiliar keepers (i.e. zoo visitors participating in the programme), relative to a control condition. Behavioural data indicate no changes in general locomotion or resting alert between conditions, nor in time spent in close proximity (less than one metre) to a keeper. However, differences were found with locomotion direction, with animals moving away from humans more than towards, and more time moving away from unfamiliar humans compared to familiar humans. Furthermore, animals engaged in less physical contact in the presence of familiar keepers compared to unfamiliar keepers. Combined, this indicates that the animals are able to distinguish between the two types of people. More time moving towards familiar keepers may represent confidence and lower fear, whereas more time moving away from unfamiliar keepers may represent increased avoidance and fear or caution. In addition, the finding of more physical contact with unfamiliar keepers is likely attributed to the opportunity for these people to feed and/or pet animals during the experience. The authors conclude that a positive experience with unfamiliar keepers in this way may contribute to increased confidence and lower fear around humans.

Visitor tours

Five studies have explored the impact of interactive zoo visitor tours. ‘Swim-with-dolphin’ tours comprise four of these, and although sample sizes were generally small ($n \leq 5$ for three of the studies), results suggest that these experiences do not have adverse impacts on the animals involved, in the short or and long term. Brensing and Linke (2003) established that dolphins can distinguish between different groups of visitors, primarily based on proximity observations (e.g. greater distance from adults compared to children). Kyngdon, Minot and Stafford (2003) explored dolphin behaviour before, during and after tours, finding increased use of the refuge area during and after, but little impact on other behaviours, including aggression. Similarly, Trone, Kuczaj and Solangi (2005) explored behaviour before and after tours, finding no changes in the frequency of social behaviour in the short or long term, but an increase in play behaviour after tours, indicating the potential benefits of such programmes. In the largest of these studies ($n = 22$), Samuels and Spradlin (1995) found greater occurrence of high-risk behaviours, including aggressive, submissive and sexual behaviours directed at humans or other dolphins, among animals involved in tours that were not under direct keeper control, highlighting the importance of keepers facilitating such interactions to avoid behaviours that put people or animals at risk or discomfort.

Adding to this, Szokalski, Foster and Litchfield (2013) explored the impact of visitor tours on big cats' behaviour immediately before, during and after tours, as well as at equivalent times on non-tour days. Lions ($n = 3$) and a tiger involved in a protected contact tour (whereby there is a physical barrier between animals and visitors) displayed decreased inactivity and increased feeding and pacing during the tour sessions, as well as more pacing and less inactivity on tour days. The increased feeding behaviour is consistent with feeding occurring during the tours, and the authors conclude that the pacing may be an anticipatory feeding behaviour rather than stereotypic. Further, cheetahs ($n = 3$) in a hands-on tour (whereby visitors enter the exhibit and can touch the animals) showed changes in multiple behaviours between conditions; however, pacing was significantly lower during tours. No aggressive behaviours were observed in either condition, and animals typically spent over half of the tour time in distant proximity (>5 m) to humans.

Combining the animal literature

Similar to the visitor literature, it is difficult to draw conclusions from a body of literature exploring a range of experiences with a diversity of species, which undoubtedly vary in terms of zoo and individual factors. Essentially, the literature comprises a series of case studies. Regardless, this body of literature has revealed no major detrimental impacts (e.g. aggression, stereotypic behaviour) of interactive experiences on the animals involved, based on the measures employed. Further, some of these studies suggest that such experiences might be enriching for some animals. In comparison to the general visitor impact literature, which indicates that visitors can have a negative impact on animal welfare for some species (Davey 2007; Hosey 2008), these preliminary findings are positive. This may be attributed to interactive experiences often being scheduled (increasing predictability for the animals involved), involving food (which might act as a positive reinforcer to reward interaction with people) and under the control of familiar zoo keepers (who dictate how interactions occur, control noise and can prevent adverse interactions). And, in many instances, animals have opportunities to retreat (e.g. move to areas that visitors cannot access, or access easily). Some of these factors have been known to contribute to positive animal welfare (Anderson *et al.* 2002; Morgan and Tromborg 2007). However, it may also represent a bias in the measures of welfare explored. For example, a number of studies employ general vigilance as the only category of animal behaviour that may indicate a watchful or wary response to visitors; this may ignore the more subtle behavioural indicators of anxiety or stress in animals, such as displacement activities (e.g. scratching/self-grooming, Troisi 2002).

Limitations and future directions

Despite an emerging body of evidence, some limitations of this body of literature hinder the ability to draw conclusions about the impact of interactive experiences on visitors and animals. From the visitor perspective, the majority of studies have explored visitors witnessing interactions, rather than engaging in interactions themselves. Therefore, conclusions about interactive experiences are limited and future research should prioritise a focus on other experiences. Further, all of the studies have employed unique outcome measures, relevant to the experience, zoo and species involved, making comparisons difficult. Methodologically, some additional limitations include the reliance on self-report measures to understand knowledge, attitudinal and behavioural outcomes (which indeed vary between studies, with no consistent tool for measuring these), and the collection of data at one point in time (e.g. immediately after the experience). While self-report measures are appropriate for understanding internal constructs, they can be open to biases. This is less problematic for constructs such as knowledge (e.g. through multiple choice questions), however self-reported attitudes and proposed behaviour change may be influenced by the strong pro-conservation messages often expressed in zoos (e.g. through signage). Further, while documenting actual behaviour change in people is difficult, researchers must consider creative ways to measure and document changes in behaviour rather than relying on self-reported intentions to change behaviour, if the true impact of interactive experiences is to be understood (e.g. see Swanagan 2000).

Similarly, objective measures of visitor behaviour during interactive opportunities might also reveal valuable insights into how visitors engage with these experiences, and could be paired with self-report measures. Termed 'visitor tracking' studies, methodologies could be employed from studies exploring typical zoo visits, including: duration and viewing behaviours (e.g. Davey 2006); time spent engaging with interpretive elements (Ross and Gillespie 2009); and differences in stay time and/or other at-exhibit behaviours to determine preferences for/popularity of different species (Carr 2016; Moss and Esson 2010). In addition to more objective measures, future studies should consider follow-up phases in order to understand the longer-term impacts of interactive experiences. Finally, despite some positive trends in the literature discussed, interactive experiences have the potential to instil incorrect messages about interacting with animals in natural habitats and exotic pet ownership (as discussed in Chapter 2). This does not appear to have been explored, but should be considered if zoos are to understand both the benefits and dangers of these experiences.

When considering the animal literature, there has been little replication of studies exploring individual experiences, with the exception of 'swim-with-dolphin' programmes. Further, the range of species explored for each experience type varies, sample sizes are typically small and restricted to one zoo, and impacts are often examined in the short term only. To account for the huge variability between interactions and associated variables (e.g. duration, location, time of day, number of visitors, activities visitors engage in, number of animals involved, whether feeding is involved), as well as individual animal differences (species, gender, age, rearing history, history of involvement in interactive experiences, etc.), it is clear that more research is required across the spectrum of experiences.

Perhaps the greatest limitation is that behaviour is the most common, and often only, measure of welfare. In addition, the range of behavioural categories measured is often too small or general, and subtle indicators of conflict or stress, such as displacement activities, are rarely included. As such, an important future direction will be not only increasing the volume of studies for each experience type and species, but also the inclusion of holistic outcome measures. While behaviour is arguably an inexpensive and useful tool for measuring animal welfare (Hill and Broom 2009), the inclusion of additional, non-invasive or minimally invasive methods, such as physiological measures (e.g. see Davis, Schaffner and Smith 2005) or keeper assessments (e.g. see Wielebnowski 1999), should be considered. Longer-term measures, such as reproductive behaviour and longevity can also provide important insights into animal well-being (Laule and Desmond 1998). Of course, not all of these are suitable, particularly when considering the impacts of an experience that occurs for a small portion of an animal's day, but behaviour alone is insufficient. Further, although this literature to date does not highlight detrimental impacts, Melfi (2009) argues for the importance of examining measures of optimal well-being, suggesting that an absence of poor well-being indicators is not enough to make inferences about good well-being. There is also the added problem that responses to stress may be delayed, and there may be cumulative effects of repeated handling or interactions resulting in chronic stress (e.g. as with capture myopathy, DelGiudice *et al.* 2005). Therefore, in the interest of taking a proactive approach to enhancing zoo animal well-being and welfare (Hill and Broom 2009), it is essential that future research explores how to maximise the potential for these interactions to be positive experiences for the animals. This could include research efforts to determine suitable animal characteristics for these experiences (e.g. personality studies).

In addition to these individual future directions for visitor and animal research, a more holistic approach to understanding these experiences is recommended. That is, none of the studies included in this review explored both visitor and animal perspectives and, with the exception of human–dolphin interactions, the visitor and animal literature cannot be matched for the same experience types. There are also gaps where some interactions have been ignored altogether – for example, the impact of feeding or patting/photography sessions on zoo visitors, or the latter on zoo animals. Future research would benefit from multi-disciplinary collaborations and research agendas, which draw on expertise of diverse researchers to consider both perspectives for the full range of experiences available. This will allow zoos to have a more complete understanding of how visitors respond to a range of experiences, and which animals are most suited to being involved. For example, if visitors respond equally to the behind-the-scenes tours with chimpanzees, tigers and dolphins, but chimpanzees demonstrate indicators of compromised well-being when involved, then zoos may consider not offering experiences with this species. Similarly, if visitors respond equally to feeding an animal through a fence compared to entering the enclosure and touching the animal, then the latter experiences might be redundant (other than the obvious human-focused opportunity for revenue raising). Finally, in light of the need to not only prevent harm to animals but also to ensure their optimal well-being in captive environments, there needs to be more equal emphasis on how animals may benefit from these experiences if zoos are to continue to justify them.

Conclusion

This chapter has reviewed an emerging body of evidence exploring the impact of interactive zoo visitor experiences on both zoo visitors and zoo animals, with a relatively equal emphasis on both participant groups. Extending what is known about traditional, non-interactive zoo visits, this review has highlighted that such experiences can contribute to visitor knowledge gain and attitude shift, although impacts on long-term behaviour change remain unknown. Further, while the animal literature indicates an absence of negative welfare outcomes associated with participating in such tours, evidence to support benefits for zoo animals is lacking. It is clear that the field still has a way to go in order to better understand the impact of interactive experiences. The promise of this form of leisure experience may be further realised with an increase in methodologically rigorous, mixed-methods studies that explore the same experiences from both visitor and animal perspectives.

Notes

1 For the purpose of this chapter, the term 'zoo' will be used throughout to refer to any official animal-keeping facility which: houses a collection of animals, is a member of an accrediting organisation (e.g. the Zoo and Aquarium Association), and has conservation, education, welfare and/or husbandry as its main aims. This can include both zoos and aquariums (adapted from Hosey, Melfi, and Pankhurst 2009).
2 In fitting with the focus of this book, studies involving domestic animals, such as petting zoo interactions, have been excluded.

3 Faecal glucocorticoid metabolite (FGM) sampling is a non-invasive process used to examine physiological change as a measure of animal welfare. Put simply, this process investigates the concentration of stress hormones (cortisol) in animal faeces under different circumstances (e.g. introduction of an enrichment item). This is often considered a more objective measure of welfare, and can be used to complement behavioural observations. For a more detailed description of this process, see Baird *et al.* (2016).

References

Adelman, L.M., J.H. Falk, and S. James. 2000. Impact of National Aquarium in Baltimore on visitors' conservation attitudes, behavior, and knowledge. *Curator: The Museum Journal* 43 (1): 33–61.

Anderson, U.S., M. Benne, M.A. Bloomsmith, and T. Maple. 2002. Retreat space and human visitor density moderate undesirable behavior in petting zoo animals. *Journal of Applied Animal Welfare Science* 5 (2):125–137.

Anderson, U.S., A.S. Kelling, R. Pressley-Keough, M.A. Bloomsmith, and T.L. Maple. 2003. Enhancing the zoo visitor's experience by public animal training and oral interpretation at an otter exhibit. *Environment and Behavior* 35 (6): 826–841.

Baird, B.A., C.W. Kuhar, K.E. Lukas, L.A. Amendolagine, G.A. Fuller, J. Nemet, M.A. Willis, and M.W. Schook. 2016. Program animal welfare: Using behavioral and physiological measures to assess the well-being of animals used for education programs in zoos. *Applied Animal Behaviour Science* 176: 150–162.

Brensing, K., and K. Linke. 2003. Behavior of dolphins towards adults and children during swim-with-dolphin programs and towards children with disabilities during therapy sessions. *Anthrozoos* 16 (4): 315–331.

Broad, S., and B. Weiler. 1998. Captive animals and interpretation – A tale of two tiger exhibits. *Journal of Tourism Studies* 9 (1): 14–27.

Carr, N. 2016. Star attractions and damp squibs at the zoo: A study of visitor attention and animal attractiveness. *Tourism Recreation Research* 41 (3): 326–338.

Curtin, S., and K. Wilkes. 2007. Swimming with captive dolphins: Current debates and post-experience dissonance. *International Journal of Tourism Research* 9 (2): 131–146.

Davey, G. 2006. An hourly variation in zoo visitor interest: Measurement and significance for animal welfare research. *Journal of Applied Animal Welfare Science* 9 (3): 249–256.

Davey, G. 2007. Visitors' effects on the welfare of animals in the zoo: A review. *Journal of Applied Animal Welfare Science* 10 (2): 169–183.

Davis, N., C.M. Schaffner, and T.E. Smith. 2005. Evidence that zoo visitors influence HPA activity in spider monkeys (*Ateles geoffroyii rufiventris*). *Applied Animal Behaviour Science* 90 (2): 131–141.

DelGiudice, G. D., Sampson, B. A., Kuehn, D. W., Powell, M. C., and Fieberg, J. 2005. Understanding margins of safe capture, chemical immobilization, and handling of free-ranging white-tailed deer. *Wildlife Society Bulletin* 33 (2): 677–687.

Dierking, L.D., Adelman, L.M., Ogden, J., Lehnhardt, K., Miller, L., and J.M. Mellen. 2004. Using a behavior change model to document the impact of visits to Disney's Animal Kingdom: A study investigating intended conservation action. *Curator: The Museum Journal* 47 (3): 322–343.

Falk, J.H., E.M. Reinhard, C.L. Vernon, K. Bronnerkant, N.L. Deans, and J.E. Heimlich. 2007. *Why zoos and aquariums matter: Assessing the impact of a zoo visit.* Silver Springs, MD: Association of Zoos & Aquariums.

Hill, S.P., and D.M. Broom. 2009. Measuring zoo animal welfare: Theory and practice. *Zoo Biology* 28 (6): 531–544.

Hosey, G. 2008. A preliminary model of human-animal relationships in the zoo. *Applied Animal Behaviour Science* 109 (2): 105–127.

Hosey, G., V. Melfi, and S. Pankhurst. 2009. *Zoo animals: Behaviour, management, and welfare*. Oxford: Oxford University Press.

Jones, H., P.K. McGregor, H.L.A. Farmer, and K.R. Baker. 2016. The influence of visitor interaction on the behavior of captive crowned lemurs (*Eulemur coronatus*) and implications for welfare. *Zoo Biology* 35 (3): 222–227.

Kyngdon, D.J., E.O. Minot, and K.J. Stafford. 2003. Behavioural responses of captive common dolphins *Delphinus delphis* to a 'swim-with-dolphin' programme. *Applied Animal Behaviour Science* 81 (2): 163–170.

Laule, G., and T. Desmond. 1998. Positive reinforcement training as an enrichment strategy. In *Second nature: Environmental enrichment for captive felids*, edited by D.J. Shepherdson, J.D. Mellen and M. Hutchins, 302–313. Washington, DC: Smithsonian Institution Press.

Martin, R.A., and V. Melfi. 2016. A comparison of zoo animal behavior in the presence of familiar and unfamiliar people. *Journal of Applied Animal Welfare Science* 19 (3): 234–244.

Mellish, S., E.L. Pearson, B. Sanders, and C.A. Litchfield. 2016. Marine wildlife entanglement and the Seal the Loop initiative: A comparison of two free-choice learning approaches on visitor knowledge, attitudes and conservation behaviour. *International Zoo Yearbook* 50 (1): 129–154.

Miller, L.J., V. Zeigler-Hill, J. Mellen, J. Koeppel, T. Greer, T., and S. Kuczaj. 2013. Dolphin shows and interaction programs: Benefits for conservation education? *Zoo Biology* 32 (1): 45–53.

Morgan, K.N., and C.T. Tromborg. 2007. Sources of stress in captivity. *Applied Animal Behaviour Science* 102 (3): 262–302.

Moss, A., and M. Esson. 2010. Visitor interest in zoo animals and the implications for collection planning and zoo education programmes. *Zoo Biology* 29 (6): 715–731.

Myers, O.E., C.D. Saunders, and A.A. Birjulin. 2004. Emotional dimensions of watching zoo animals: An experience sampling study building on insights from psychology. *Curator* 47 (3): 299–321.

Ogle, B. 2016. Value of guest interaction in touch pools at public aquariums. *Universal Journal of Management* 4 (2): 59–63.

Orban, D.A., J.M. Siegford, and R.J. Snider. 2016. Effects of guest feeding programs on captive giraffe behavior. *Zoo Biology* 35 (2): 157–166.

Pearson, E.L., R. Lowry, J. Dorrian, and C.A. Litchfield. 2014. Evaluating the conservation impact of an innovative zoo-based educational campaign: 'Don't palm us off' for orang-utan conservation. *Zoo Biology* 33 (3): 184–196.

Povey, K.D., and J. Rios. 2005. Using interpretive animals to deliver affective messages in zoos. *Journal of Interpretation Research* 7 (2): 19–29.

Price, A., E.R. Boeving, M.A. Shender, and S.R. Ross. 2015. Understanding the effectiveness of demonstration programs. *Journal of Museum Education* 40 (1): 46–54.

Ross, S.R., and K.L. Gillespie. 2009. Influences on visitor behavior at a modern immersive zoo exhibit. *Zoo Biology* 28 (5): 462–472.

Samuels, A., and T.R. Spradlin. 1995. Quantitative behavioral study of bottlenose dolphins in swim-with-dolphin programs in the United States. *Marine Mammal Science* 11 (4): 520–544.

Sherwen, S.L., P.H. Hemsworth, K.L. Butler, K.V. Fanson, and M.J.L. Margrath. 2015. Impacts of visitor number on kangaroos housed in free-range exhibits. *Zoo Biology* 34 (4): 287–295.

Smith, L., S. Broad, and B. Weiler. 2008. A closer examination of the impact of zoo visits on visitor behaviour. *Journal of Sustainable Tourism* 16 (5): 544–562.

Swanagan, J.S. 2000. Factors influencing zoo visitors' conservation attitudes and behavior. *Journal of Environmental Education* 31 (4): 26–31.

Szokalski, M.S., W.K. Foster, and C.A. Litchfield. 2013. Behavioral monitoring of big cats involved in 'behind-the-scenes' zoo visitor tours. *International Journal of Comparative Psychology* 26 (1): 83–104.

Troisi, A. 2002. Displacement activities as a behavioral measure of stress in nonhuman primates and human subjects. *Stress* 5 (1): 47–54.

Trone, M., S. Kuczaj, and M. Solangi. 2005. Does participation in dolphin-human interaction programs affect bottlenose dolphin behaviour? *Applied Animal Behaviour Science* 93 (3): 363–374.

Visscher, N.C., R. Snider, and G. Vander Stoep. 2009. Comparative analysis of knowledge gain between interpretive and fact-only presentations at an animal training session: An exploratory study. *Zoo Biology* 28 (5): 488–495.

Weiler, B., and L. Smith. 2009. Does more interpretation lead to greater outcomes? An assessment of the impacts of multiple layers of interpretation in a zoo context. *Journal of Sustainable Tourism* 17 (1): 91–105.

Wielebnowski, N.C. 1999. Behavioral differences as predictors of breeding success in captive cheetahs. *Zoo Biology* 18 (4): 335–349.

4 Consumer perceptions of keeping wild animals in captivity

Susanna Curtin and Eleanor Green

Introduction

Viewing animals in captivity and taking part in animal-based activities has become a regular part of many people's tourist experience. Recent research suggests that wildlife or animal-related tourism accounts for 20–40% of global tourism (Moorhouse *et al.* 2015). A poll conducted in 2015 by World Animal Protection found that 93% of tourists choose to take part in animal-based tourism just because they 'love' animals or simply to 'have fun'. According to Daniel Turner of the Born Free Foundation (personal communication 2016), approximately 70% of tourist excursions involve animals.

Fascination with the animal kingdom is an ancient, universal and cross-cultural phenomenon. Our evolution has been inextricably bound by our relationship with them and animal imagery has long inhabited the human imagination (Rothfels 2002). From the cave paintings, stories and rituals of ancient civilisations to the modern adaptation of children's stories, films, television programmes and the keeping of pets, animals co-inhabit our everyday and we draw inspiration from their beauty, complexity and diversity. In today's societies where we have become more separated from the natural world, there lurks an inherent desire to gaze upon and interact with the animal-other in a safe, contrived and commodified environment (Cunningham 2016). It is therefore no wonder that animals are a mainstay of global leisure and tourism experiences, and that the spectrum of animal-tourist experiences is growing. Opportunities to experience wild animals exist in both captive and non-captive settings. Non-captive settings are where we can watch free-ranging wildlife in their natural habitat (Newsome *et al.* 2005), whereas captive settings are tourist attractions which include zoos and safari parks, marine parks, elephant parks, bear parks, tiger interactions and lion encounters, sanctuaries, farmed wildlife attractions (civet coffee, sea turtles, tigers, crocodiles, bear bile), street performances (macaques, hyenas, snakes, bears), animal rides (horses, donkeys, elephants and camels) and domesticated farm animals (Moorhouse *et al.* 2015).

Out of this list, zoos represent the most popular experience given that visiting them features strongly in school children's itineraries and family entertainment (Therkelsen and Lottrup 2015). Jamieson (2006, 132) defines zoos as "public parks which display animals primarily for the purpose of recreation and education". It is impossible to measure the precise number of zoo visitors throughout the world, however based upon their 1,200 members, the World Association of Zoos and Aquariums (WAZA 2016) report over 700 million visitors annually. As a result of this huge demand for animals in tourism there are now over 100 million captive animals working for the purpose of entertainment (Fennell 2013).

There have been a number of recent incidents in the press which question the future of zoos. The shooting of Harambe the gorilla at Cincinnati Zoo after a four-year old boy 'fell' into the enclosure, and the public slaughter of Marius the young, surplus giraffe who was fed to the lions in Copenhagen Zoo has reignited the ethical debates of keeping animals in captivity for our own entertainment, as well as the scepticism of zoos' true conservation value. The question is whether captive animal-based tourism attractions such as zoos should have any place in our changing society where there is increasing interest in animal welfare (Doward 2016; Finch 2008). This chapter will investigate this question by exploring the views of consumers who have visited a zoo or a similar captive animal attraction in 2016. It will investigate the appeal and fascination of animal-based attractions, the attitudes of tourists towards animal welfare within tourism and the educational and conservational influence of animal-based attractions on tourists, thus being able to suggest the potential role of these attractions in the future.

Were it not for these attractions most people would never get the chance to safely view the diverse, exotic and captivating animal kingdom with whom we share our planet. Despite his dislike of the captive animal industry, Croke (1997, 14) eloquently proposes that in a world increasingly disconnected from nature "the zoo provides a venue for us to link souls with wildness", allowing us to re-engage with a number of iconic and charismatic species in a safe, contrived setting. Other ways of legitimising zoos include the assumption that the captive wildlife setting provides important conservation programmes, either in terms of protecting diminishing species, reintroduction programmes and/or educating us on animal behaviour and threats to their survival. Regardless of the good work captive animal attractions might do, the concept of captivity has sparked considerable debate in recent decades (Woods 2002). The ethics of keeping animals in unnatural settings and enclosures where they cannot display their normal behaviour is becoming more and more questionable as our knowledge of animal behaviour and their suffering in captivity increases (Ferdowsian *et al.* 2011; Hutchins *et al.* 2008).

Notions of captivity

The word captivity is loaded with negative connotations in a human dialogue. It represents confinement, imprisonment, slavery and incarceration, and epitomises the very worst human existence where all freedoms are curtailed. It is somewhat surprising then how accepting we are of animal captivity; what is it that prevents us from extending these values to the animal other? It could be argued that our need and desire to be close to animals overrides our sense of moral duty. Wood (2002) suggests that being close to wild animals, interacting with them, learning about them and seeing such a vast variety of animals and animal behaviour evokes memorable and emotional experiences which makes us blissfully unaware of the suffering animals endure to be the subject of our amusement (Moorhouse *et al.* 2015). A good example of this is illustrated by the SeaWorld experience. Who could ever imagine how dolphins, a creature used to belonging in large social groups and covering vast tracks of the world's oceans could ever feel at home in a water tank less than the size of the average car park, and yet many people still flock to these venues for their entertainment even after the increased publicity of the dark realities exposed in the film Blackfish. Indeed, attendance numbers at SeaWorld Entertainment's US parks rose by 5.6% in 2014 to reach 3.2 million visitors (Walters 2015). This reveals a dichotomy in the zoo/aquaria market between those who see animals as having moral rights and those who purely focus on their own and their family's entertainment.

The capture and keeping of animals is clearly not a new concept and has been prevalent throughout human history (Fennell 2012). A fascination for exotic creatures began in the exploration and conquest of new lands, and menageries sprang up in nations to illustrate power and wealth. Caged wild animals were given as gifts by royalty, and many noble families had their own private collection of animals. As cities developed, the display and viewing of captive animals became an art form to connect people to the wild (Veltre 1996).

Fennell (2012) argues that the origins of the modern zoo began in the Age of Enlightenment, which was a cultural movement of intellectuals in the 17th and 18th centuries following in the wake of philosophers such as Descartes who espoused the use of animals for our growth and entertainment. Animal shows took to the road accompanied by a knowledgeable compere who informed the public of the geography and biology of the species whilst preaching Descartes' supposition of the irrational nature of animals in comparison to the rational nature of man (man versus beast). These travelling menageries opened the door for the public acceptance of animal confinement and zoos as an accepted cultural institution. From here zoos developed from a menagerie of living natural history in the 19th century, to a zoological park (i.e. a living museum) in the 20th century and then to a centre of conservation in the 21st century (Rabb and Saunders 2005), depicting a journey which corresponds to public demand, curiosity and entertainment rather than animal welfare.

'Zoo' enclosures

In the 19th century, zoos kept animals behind metal bars in concrete enclosures with no natural habitat (Frost 2011; Hughes *et al.* 2005). These enclosures offered constant exposure to visitors and noise from the outside as well as a threat of violence from usually unqualified keepers (Marino *et al.* 2009). According to Marino *et al.*, many of the captive animals would display behaviours and emotional states that reflected their psychological trauma and distress. They would purposefully injure themselves, act depressed and show hyper-aggression; behaviours that today are recognised symptoms of 'zoochosis'. Distressed animals are not pleasurable to behold and therefore these animal welfare issues affected the public's consciousness, and zoo visitor numbers declined between the 1950s and 1970s (Frost 2011). For example, the number of visitors to London Zoo plummeted after reaching its peak in the 1950s.

After concrete enclosures and metal bars became frowned upon by the public, responsible zoos modified their animal enclosures. Invisible barriers and sunken enclosures surrounded by moats replaced metal bars (Frost 2011; Tribe 2004). In addition, more vegetation and aspects of natural habitats were carefully placed within the enclosures, giving a more natural appearance. After these visual changes there was a call to improve captive animals' experience by providing more things for them to do ('behavioural enrichment programmes'), such as allowing them to search for food, have places to hide and more 'play' things in their enclosures in an attempt to reduce zoochotic behaviours. In response to public demand and greater public awareness of the need for conservation given the destruction of so many natural habitats, there also came the promise of greater conservation effort and more emphasis on visitor interpretation within zoos (Frost 2011).

Widespread public concern for the protection of ecosystems and animals prompted further changes to reflect the conservation needs of species in the wild (Frost 2011). This made zoos more desirable to the public and they gained financially from the increased visitation (Tribe 2004). The majority of modern zoos now have five primary goals: animal welfare, conservation, education of the public, research and entertainment (Fernandez *et al.* 2009), with natural enclosures enhancing visitor satisfaction. However, perceptions of enclosures can be deceiving for the average visitor. The size of an enclosure often remains the same following this 'gentrification'; visual barriers and bars are dismantled and replaced by moats and ditches, giving the illusion of greater space, freedom and welfare, but this is more for the public's benefit than the animals (Born Free Foundation, personal communication, April 2016). Space should ideally be in relation to the size of the animal and how they use space in the wild (i.e. their home range or territory) (Young 2014).

Setting animal welfare regulations and standards

Fennell (2012) reminds us that not all zoos or captive animal attractions are managed the same. For example, there are good examples and bad examples with many in the developing countries providing sub-standard living conditions for their animals. Moreover, despite the European Union (EU) Zoos Directive, only about one third of the 1,000 zoos in the EU were found to be reputable enough to include in the International Zoo yearbook, and there are many countries such as Greece, Portugal, Spain and Italy where there are no laws to govern the setting up of animal attractions for public entertainment. Figures 4.1 and 4.2, for example, are photographs taken in Bulgaria and Serbia in 2016; countries governed by EU directives but which lack regulatory bodies. Here there is little to stop anyone setting up a zoo and keeping animals in unsuitable enclosures and little enforcement of EU laws in rating the performance of zoos against their

Figure 4.1 Lion in a Bulgarian zoo (photograph courtesy of the Born Free Foundation 2016).

Figure 4.2 Bear in a Serbian zoo (photograph courtesy of the Born Free Foundation 2016).

legal requirements. In their 2011 report for the EU Zoo enquiry, the Born Free Foundation reported that housing conditions for the zoo animals in Bulgaria are "generally poor, run-down and restrictive" (p. 14).

In 2013 the Association of British Travel Agents (ABTA) released their first Global Welfare Guidance document for Animals in Tourism (GWGAT). This was the first significant regulatory tourism body to propose welfare guidance for all captive animal-based attractions worldwide. ABTA's Animal Welfare Vision (2013) sets out five key objectives for the industry:

1 to understand the scope of tourism's impact on animals;
2 to raise awareness of animal welfare best practice with governments, suppliers and customers;
3 to help members develop means to assess and improve performance within the tourism supply chain;
4 to review and report on actions;
5 to set targets for long-term improvement.

It also provides six supporting guidance manuals. These manuals are a benchmark for "best practice in animal welfare" within tourism and captive animal attractions worldwide and it is hoped that, with cooperation from ABTA industry members and destination governments, fundamentals of bad animal welfare practice will eventually be reduced and/or phased out. The ABTA animal welfare manuals are: *Animals in Captive Environments; Dolphins in Captive Environments; Elephants in Captive Environments; Wildlife Viewing; Working Animals;* and *Unacceptable and Discouraged Practices.*

Traditionally, animal welfare has revolved around the philosophy of the Five Freedoms (initially devised by the UK Farm Animal Welfare Council 1979). These are: (1) freedom from thirst, hunger and malnutrition; (2) freedom from discomfort and exposure; (3) freedom from pain, injury and disease; (4) freedom to express normal behaviour; and (5) freedom from fear and distress (see Table 4.1).

Table 4.1 The Five Freedoms

Freedom	*Definition*
Good feeding	Absence of prolonged hunger Absence of prolonged thirst
Good housing	Comfort while resting Thermal comfort Ease of movement
Good health	Absence of injuries Absence of disease Absence of pain induced by inappropriate management procedures
Appropriate behaviour	Expression of social behaviour Expression of natural behaviours Good human–animal relationship Positive emotional state
Protection from fear and distress	Absence of general fear / distress / apathy Ability to seek private refuge Absence of surgical or physical modification of the skin, tissues, teeth or bone structure other than for the purposes of genuine medical treatment / manipulation / sedation

The Five Freedoms are still a highly influential framework as they direct attention towards the need to understand, identify and minimise negative welfare states. Whilst they undoubtedly set a benchmark, they are not without criticism for their simplicity; particularly that they do not "capture, either in the specifics or the generality of their expression, the breadth and depth of current knowledge of the biological processes that are germane to understanding animal welfare and to guiding its management" (Mellor 2016, 21).

In addition, current laws and licenses only set minimum standards for animal management (Captive Animal Protection Society 2016). Mellor (2016) espouses that for animals to have "lives worth living" it is not just about minimising negative experiences but about optimising positive ones. In addition, Melfi (2009) believes that existing efforts to improve the welfare of captive animals are biased to very few species and that for success to be achieved, welfare regulation should be part of national legislation and be species-specific. As Webster (1995) proposes, it matters not how we think or feel towards issues surrounding animals, but what we actually do. As consumers, what we decide is often based on knowledge or what we see and sometimes it is difficult to see the reality.

Hidden realities of captivity

Regardless of individual attitudes towards animals, World Animal Protection (2015) predicts that 80% of tourists cannot see the negative impact that animal-based attractions have on animal welfare. Animal welfare is loosely defined by Dawkins (2006) as a two-step approach: (1) if animals have what they want and

Figure 4.3 A solitary zoo gorilla (photograph courtesy of the Born Free Foundation 2016).

Figure 4.4 Tiger in a Romanian zoo (photograph courtesy of the Born Free Foundation 2016).

(2) if they are happy. This brings psychological and behavioural measures into the assessment of welfare but is both limiting and unmeasurable. How can we be sure that an animal is 'happy'? It is probably easier to determine when they may be 'unhappy'. For example, in Figures 4.3 and 4.4, the animals are in very small enclosures. There is no enrichment to alleviate boredom. There is nowhere to hide or escape from visitors, and in the case of Figure 4.3, they are solitary and away from their normal social group. Indeed, the more science reveals about the animal kingdom, the more captivity is questionable given how it may, although not always, shorten life expectancy (Robeck *et al.* 2015) and cause psychological imbalances (Mason 2006; Mason and Rushen 2008). Although some animals are known to adapt better than others, Knight (2009) argues that it is immoral to keep wild animals in enclosures a fraction of the size of their natural habitat, forcing them to be on display in close proximity to and/or performing for tourists. Many animals show signs of distress, even when bred in captivity, and most will try to escape at any given opportunity from their enclosures (Mason and McCarthy 1995).

According to Hediger (1964) the limitation of space prevents an animal's normal flight reaction. They cannot limit their territory and therefore they cannot prevent encounters with animals of their own species, or protect themselves from us as their ultimate predator. We have all witnessed the pacing of animals in confinement as they attempt to cover the physical distances they would in the wild (Breton and Barrot 2014) or deal with the boredom that an enclosure instils. There is also increasing scientific evidence and recognition that animals have feelings and are sentient beings (Bethell 2015) and therefore will feel the effects of confinement.

Elephants, in particular, have attracted a significant number of studies as they are known to suffer intensively from performing tourist activities and from post-traumatic stress suffered as a result of being taken away from their family groups (Cohn 2006) or by sustaining the brutality of their keepers in many Asian attractions as Turesson (2014, 5) explains:

> The story of a baby elephant that has been captured by humans usually goes like this: In a village in Thailand there is an elephant calf in a wooden cage. She is about four years old and has just watched her mother being killed when she tried to protect her baby from the people that now have her. She was illegally taken with brutality from the wild. Now she is immobilized in this small cage and there are people stabbing her with sharp tools over and over again. This beating goes on for days, with no food or water, because these people are using an old Asian ritual to break the elephant's spirit. They want to make it fear humans so it will obey them. This is the reality for most of the elephants living in captivity in Asia, not just in Thailand.

(See also King 2005; Kontogeorgopoulos 2009; Duffy and Moore 2011; Thongma and Guntor 2011.)

An academic journal is wholly dedicated to applied animal welfare science (*Journal of Applied Animal Welfare Science*), which includes studies of the effect of captivity on animal welfare. The current focus of the studies in this journal is on captive animals' daily activity budgets, the time spent pacing or exhibiting other stereotypic behaviour, the effects of different housing on large mammals, and how the density and intensity of visitors affect animal welfare. The purpose of these studies is to better understand how animals respond to captivity and how best to improve their environments. They also question the role of zoos and the conservation that they do, and highlight major welfare problems in particular countries. They challenge zoos to address the changes required and governments to improve legislation.

Tourist attitudes towards captivity: complexity and confusion

There is some controversy in the literature with regards to how the paying public view animal welfare, with some suggesting that western consumers are becoming more aware of their moral responsibilities with regards to captivity whilst others suggest otherwise. Phillips (2009) asserts that the paying public are much more environmentally aware now than ever before and attitudes towards animals are slowly shifting, with the perception of animal welfare being influenced by a progressive society and new generations. Likewise, MacQueen (2009) proposes that despite there being a significant demand for captive animal attractions, people's attitudes and values towards these captive animals have noticeably shifted. The outdated mechanical view of animals is now being challenged by modern interpretations regarding the conceptualisation of animals as subjects rather than objects and by implication, the moral responsibility facing humans when dealing with them (Bertella 2014; Armstrong and Botzler 2008). This sentiment is evidenced by a YouGov Poll in the United Kingdom conducted in 2012 based on a sample of men and women aged from 18 to 55 and across various social groups. The survey indicated that alongside a strong tourist demand to see or interact with animals, tourists want to be assured of good animal welfare standards. Over half of British adults say that if they were on holiday abroad and saw an animal being mistreated it would put them off visiting that country again. Indeed, Shani and Pizam (2008) suggest that tourists are unlikely to condone keeping animals in poor condition.

Nonetheless, whilst these caring attitudes are prevalent, they may not be carried over into behaviour. In the 2012 YouGov Poll, just over one in five British adults (22%) had seen animals being mistreated when on holiday overseas, yet over three-quarters of these holidaymakers (77%) made no attempt to report the most recent incident of mistreatment that they saw. Moreover, the evaluation of 188 TripAdvisor posts with regards to specific animal attractions conducted by Moorhouse *et al.* (2015, 1) reveals that even when the welfare of the animals is poor, 80% of tourist reviews were positive. They conclude that "substantial negative effects are unrecognised by the majority of tourists, suggesting an urgent need for tourist education and regulation". Shani (2012) concludes that the extent to which tourists agree with the existence of an animal-based attraction stems from how well animals' captivity can be justified. For example, conservation, protection and education are rated the most persuasive justifications for captivity but just how believable are these today? Previous studies have questioned these justifications (Ballantyne *et al.* 2009; Balmford *et al.* 2007; Smith and Broad 2008; Bulbeck 2005), suggesting that for the visitor, captive animal attractions such as zoos are about recreation and enjoyment rather than education. Even the 'conservation effort' that is promised by zoos is debated by lobbyists such as Peter Singer (2016).

Recent events such as the culling of Marius the unwanted giraffe in Copenhagen Zoo (see Cohen and Fennell 2016), the exposure of SeaWorld (Neate 2016), the gorilla incident in Cincinnati (Oluo 2016) and the death of a tourist killed on an elephant trek in Thailand (Quinn 2016) expose the truth and disclose the ethical issues of the use of animals for our entertainment. These stories and others have catapulted issues of animal welfare within tourism into the limelight and contributed to a sudden rise in media attention potentially causing a global stirring of changing views. In order to explore these attitudes we undertook a small research project with some UK citizens to explore their attitudes towards captive animal attractions.

Methods

Our research methods included two focus groups; one consisting of parents (whose children were under 15) and the other consisting of young people aged 20–30 without children. Given the recent media stories of animals in captivity, focus groups were selected due to their ability to elicit rich data provoked by a dynamic discussion that group interviews allow. The groups were held in February 2016. Social networks were used to select participants who must have visited a zoo in the last 12 months. The discussion guide incorporated questions regarding motivations for visiting, awareness of conservation efforts, perceptions of animals in captivity and the entertainment that zoo animals provide. Discussions were recorded, transcribed and analysed using a thematic analysis.

To enhance this data, an online survey was undertaken using SurveyMonkey in March 2016. This comprised 164 respondents who had visited an animal attraction in the last 12 months. The questionnaire consisted of open and closed questions and was designed to collect and measure people's attitudes and opinions towards captive animal attractions (including zoos). Themes included the appeal of captive animal-based attractions, feelings towards captive animals, animal welfare, the key roles of captivity and the influence of media stories. Data was analysed using SPSS.

Exploring visitor motivations

An important objective in conducting the online questionnaire was to assess peoples' motivations for visiting captive animal-based attractions. The data reveals that 74% of respondents' primary motivation was "to have an enjoyable family day out", 68% of respondents wanted to see species that they would not otherwise get to see and 40% indicated they wished to see animals in captivity in order to learn about them. This supports previous research by Roe *et al.* (2014) who found that visitors mostly chose to visit a zoo for a novel day out, and not all perceive this as an opportunity to learn about animals. Research by Frost (2011) also indicated that the majority of zoo visitors go to be entertained.

Two themes emerged from the focus groups. The first was entertainment for children. Participants indicated that zoo visiting is often instigated by "having spare time to fill with children" and the notion that "zoos are good entertainment". For parents of young children (a key market for animal attractions, Mintel 2014) it was also about sharing nice, memorable days out with them and "for our children to appreciate the 'real' animals that are depicted in their story books". It is somewhat ironic that seeing captured animals is deemed child entertainment given the distress that animals in captivity can feel. The other prominent theme concerned the "fulfilling of lifelong ambitions". This was particularly apparent in the case of more exotic animal attractions such as riding or interacting with elephants, swimming with dolphins or walking with lions, and is further evidence of a utilitarian and anthropocentric view of the animal kingdom, existing in order fulfil our dreams.

While the captivity of any living creature is questionable, it is the large charismatic mammals that have attracted the most research on behaviour in captivity (Robeck *et al.* 2015; Mason 2006). This is partly in recognition that large mammals in the wild often travel large distances and therefore do not adapt to captivity as well as other species which have a smaller range (Cohn 2006). The problem, however, for zoos is that it is these large charismatic animals that attract the most attention from visitors are the most appealing: as one of our participants said: "people want to see a panda, lion or elephant; they don't necessarily want to see a tarantula or a snake". Recent research by Carr (2016) reveals a similar story that large mammals tend to dominate tourist motivations. Other desirable traits found in his sample of zoo visitors included endangered, active animals and those who display intelligence (see also Barney *et al.* 2005 and Christie 2007).

Perceptions of captivity

When respondents of our online survey were asked about how they felt seeing animals in captivity, 40% said that it was dependent on the animal and the enclosure: "if they seem happy and well cared for with adequate space I feel it is a rewarding experience, however, if they are poorly cared for or in spaces too small for them I feel angry and sick". Feelings of sadness and anger were apparent (34%):

> it saddens me greatly. More and more I feel the line between humans and other species is blurred. When I see elephants in chains I see them as suffering like slaves … It appals me that we have not evolved further in the way we treat and perceive animals.

Guilt was also expressed by 15% of respondents: "I shouldn't be there/endorsing this" and "why did I pay the entrance fee…?" and a further 11% reveal mixed emotions including awe and/or excitement. Overall, respondents struggle with the concept of captivity, with feelings such as guilt, sadness and anger. They are aware of a 'right' and 'wrong' and hold captivity as an unfavourable choice. But they still go to see captive animals, thereby supporting the very thing they question.

The findings of our study reveal value-conflicts on several levels, the first being the inherent dislike of seeing animals in captivity being over-ridden by an intense desire to see them (see Demello 2014). In the case of zoo visitors, most participants voiced similar feelings of guilt or unease about captivity and nearly all looked for ways to reduce the dissonance they felt. Their disquiet often had to do with the perception of space and enrichment, with some relating scathing stories about the zoos they had visited. Other participants show concern that some animals should not be kept in the British climate, especially if they derive from exotic countries. Anthropomorphic tendencies make respondents more receptive to feeling empathy towards the animals that are living away from their natural habitat and climate. This point is reiterated by People for the Ethical Treatment of Animals (PETA) (2016) who also argue that exotic animals should not be kept in a climate unsuitable for them.

A second value-conflict was the manner in which animals that are familiar, or apparently so, in our everyday life were deemed less of a worry in terms of welfare and conservation and were dismissed within both focus groups as animals worth discussing or indeed viewing: "I think the difference is a Capybara just looks like a strange sheep or … something and so it's in an enclosure and you're used to seeing things like that in enclosures and so it doesn't seem so bad"; "See if it [Capybara] looks like a sheep I wouldn't even stop or pay attention to that because I'm thinking it's just a sheep, even though it's not". These comments suggest that we do not feel there is an issue with viewing animals which are perceived to be related to domesticated species; it is more of an issue if the animal is exotic. This stresses the complexity of the human–animal encounter and highlights how attraction and empathy is species driven (Sommer 2008), something highly evident in conservation where it is difficult to rally support, interest and funding for non-charismatic taxa such as crocodiles and sharks.

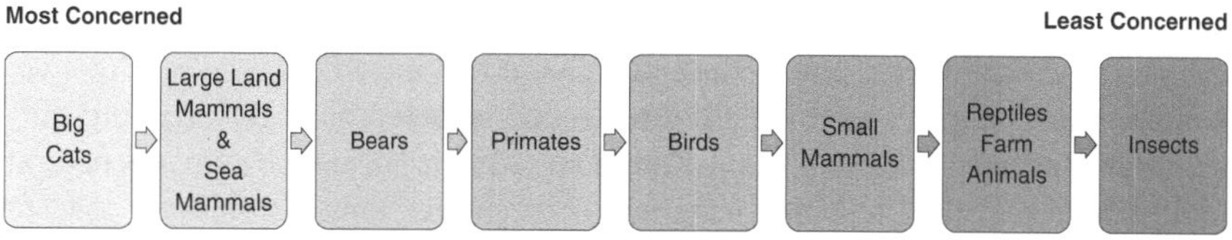

Figure 4.5 Spectrum of concern.

Focus group participants were asked to rank the types of animals they might 'feel sorry for' in captivity. Figure 4.5 shows this spectrum of concern which, once again, replicates the significance of large charismatic mammals and primates with reptiles and insects evoking the least empathy.

Most participants acknowledge that seeing conservation taking place within the zoo made them feel less guilty for viewing the animals: "if they've got something there, that if it wasn't in the zoo it would be dying out, then I don't feel so bad about standing there viewing it". The problem is that the average visitor may not know which animals are 'dying out'. As one participant reveals:

> I think it [captivity] is OK if I know they are super extinct, so … for example, I don't think tigers are that extinct in relation to other species that are literally on the brink of extinction, then I'd feel oh good because they're almost gone, so I'd feel much happier than I would for a species that's not so extinct.

Quotations such as this reveal that visitors may not be well educated in the plight of animal conservation as tigers are on the International Union for the Conservation of Nature red list of threatened species due to the encroachment of humanity, loss of habitat and hunting (WWF 2016).

Beardsworth and Bryman (2001) refer to the 'Disneyization' of zoos. They explain how emotional labour is used to induce feelings of guilt in terms of the wider environment and species extinction, and how zoos then offer conservation as justification for their activities and captivity as a solution. This brings us on to the next conundrum: although the zoo participants talk about conservation sweetening the notion of captivity, their responses to questions about the conservation that zoos undertake reveals that they are unconvinced by captive breeding programmes or re-release: "It's all well and good breeding animals in captivity, but they're breeding them into captivity and having them live in captivity again"; "I don't like … it's nice that they're re-breeding them, but for what? That's just getting people through the doors to go and see a baby Rhino isn't it?".

Likewise:

> There is a lot of emphasis on conservation and breeding programmes, but I think a lot of the zoos, they don't do it for the sake of conservation, because their motive isn't to release them back into the wild, they breed them to create more money at the end of the day, so to keep the cycle of visitation.

Dolmage (2014) argues that zoos often use breeding programmes as "their good deed to conservation". In reality a zoo is the last place a captive breeding programme with the intent to reintroduce the animal into the wild would be effective. It is common practice for animals to be traded between zoos for breeding purposes which disrupts social groups. Zoos also fail to replicate the animal's natural environment and tend to breed generations of captive animals who are used to humans, none of which is conducive to successfully reintroducing them into the wild. However, breeding programmes create new baby animals, which are highly valued by zoos as they lead to increased visitation. Pekarik (2004) claims that baby animals enhance anthropomorphic tendencies in human audiences which connect to our nurturing instinct. As well as being cute, they also enhance the possibilities for metaphorical thinking and open up the possibilities for families to discuss deeper matters, such as reproduction, with their children.

The way zoos market new baby animals as a unique selling point also creates some cynicism with our participants:

> If somewhere has got a baby, like a snow leopard or something ... they advertise it and everybody wants to go and see it ... and then I don't know ... is it because of the conservation of snow leopards or simply because "oh look it's a really cute baby snow leopard" which gets new people through the gates.

Conservation of species, although lessening guilt, is not the respondents' main motivation for visiting. Furthermore, some participants feel that the conservation within zoos is misplaced: "I think that true conservation is more like monitoring them in their natural habitat rather than nurturing them in a zoo environment". This is a sentiment echoed by the World Association of Zoos and Aquariums (WAZA) (2005), an industry group of 280 zoos and aquaria, which urge zoos to reconsider their captive breeding programmes and to consider more habitat conservation and genuine reintroduction efforts instead.

Education of visitors

Fennell (2013) claims that due to the growing stigma attached to zoos for having entertainment as their primary goal, they are now focusing on educating the public on conservation, habitat protection and the natural environment. Therefore, around zoos there are many plaques with information to educate visitors about the animals and their plight in the natural world. However, unanimously both focus groups argued that these snippets of information were "boring" and "dry" and that there must be better ways of trying to educate the public about the animals: "I think they could make a, be a bit more creative in the way they do it [interpretation], there is always dry postcards, and you've got to stand there for five minutes to read it"; "I think urm, they have a lot of things up on the wall, but whether people actually read them is another matter you might to begin with but then you get a bit bored and you're like I don't really care that much anymore".

This is synonymous with Demello's research (2014), which found that most visitors do not read the information attached to enclosures and this can indicate that little education is taking place within the zoo. The only exception, according to De Koff (1998), is when the information is provided in a more interactive manner with keeper presentations and crowd involvement. So if education of visitors is limited and conservation efforts questionable, the entertainment factor of animal attractions is clearly the main focus of a visit.

Animals as entertainment

In our online survey, 70% of respondents had interacted with animals at some point during their vacation, either feeding, holding or petting them; 37% had attended animal performances; and 31% had taken part in animal rides. Despite their having taken part in these activities, 77% of them disagreed with animal performances: "animals shouldn't be used for entertainment … we do not need to dominate other animals for pleasure and funding violent industries that train animals for these purposes is unethical", and 72% disagreed with animal rides: "I think that rides are potentially more questionable … as they require a greater degree of animal training … have we reduced elephants to facilitators of exotic donkey rides" and "animals do not deserve straining treatment". Interestingly, petting, feeding and close interaction with animals incurred less controversy (even though in reality animals often undergo teeth removal or sedation in order to enable handling, Born Free Foundation 2016). Two factors seem to be at work here. The first is that reaching out and touching is paramount to establishing and keeping social contacts (see Konecki 2008) and that this instinctive human behaviour to connect with other living things is then carried over into our contact with domesticated or appealing animals. The second is that nearly half of UK households own a pet (PFMA 2015) and that animals used for petting or close interaction create a sense of familiarity, replicating the closest and most humanised of human–animal relations (Franklin 1999).

On reflection, group members admitted to "feeling a little sad" following a visit to a zoo or other animal attraction as there was usually at least one exhibit or experience that made them feel this way: "I think from an adult's point of view you do feel sad, you feel like …. I've just given my money to … to see other animals in captivity, you know, it's like a lesser life as such"; "it isn't their home – it is not where they belong".

Although some participants acknowledged the convenience of zoos and paying one price to see lots of animals that they could not afford to see in the wild, the notion of profit and the commercial focus created further dissonance as the following quotations depict:

- "For me … it's where does the surplus funds go? Does it go back into conservation and breeding programs in the wild or does it actually go into building a restaurant or profits for the chief executive".
- "It's always going to be about money though isn't it at the end of everything it's always money".
- "It's about the greed'.
- "They just really use the animals for the sake of the profits".

Conclusion

Key findings reveal the strong dissonance that tourists feel towards captive animal attractions: They are morally torn but aesthetically persuaded. On the one hand, they see animal attractions as interesting, culturally significant and enjoyable for everyone in the family, but on the other, feel discomforted by the notion of captivity and large creatures in small places. The promise of the conservation of species negates some negative emotions, although when probed further, participants from the focus groups exposed their mistrust of the conservation ethic and subsequent captive breeding programmes.

Although from a small-scale study, our findings highlight two major outcomes. First, that the British public are beginning to question the morality of keeping animals captive for their enjoyment and entertainment but that they still see visiting animal attractions as an enjoyable and memorable family day out, seeing animals that they cannot see elsewhere. The conundrum is that these two values do not sit very comfortably and so in order to reduce dissonance, consumers look for ways to justify buying their tickets. The second finding is that the conservation that zoos supposedly carry out helps the consumer to counter their disquiet. Yet, at the same time, they do not totally 'buy into' the type of conservation that zoos purport to take part in and doubt the conservation impact that they are having. These inner conflicts suggest that consumer behaviour does not mirror their attitudes; they still buy the ticket. There is a need for further research into this disconnect and to find ways to challenge and change social acceptance and behaviour towards animals in captivity given the fact that people are inherently questioning the confinement of intelligent animals for our enjoyment.

We believe that zoos will eventually need to change in order to remain attractive. That they should be far more transparent regarding their breeding and disposal of zoo animals, that they should only house and exhibit species which are known to thrive in captivity, and finally, that they should publicise the fact that large mammals such as elephants do not thrive in captivity and that is why they no longer house them. Keulartz (2015a, 349) concludes that

> the zoo is standing at a crossroads; it has to decide if it will fully commit to the new paradigm and develop into a true conservation centre or if it will degenerate (further) into a venue for entertainment that will provoke increasing criticism, not only from animal protectionist but also from wildlife conservationists.

Peter Singer, one of the first ambassadors of the animal rights movement, was recently asked whether zoos should exist for the sake of species preservation; he replied that animal attractions confine animals for our amusement in ways that are contrary to their interests. Even if these zoos do occasionally preserve an endangered species, "what is the point of preserving animals if they are having miserable lives?" (cited Keulartz 2015b, 254), a sentiment also proffered by Jamieson (1995). It remains, however, that lots of animals are still living these lives in order to provide the paying public with memorable tourist experiences and days out; a simple search of the internet reveals countless examples. But whilst consumers continue to buy the ticket the practice of animal exploitation will continue. We should know better by now.

References

ABTA (Association of British Travel Agents) (2013) *ABTA Animal Welfare Guidelines*. Available from: https://abta.com/abta-shop/abta-animal-welfare-guidelines [Accessed 20 November 2017].

Armstrong, S.J. and. Botzler, R.G. (eds) (2008) *The Animal Ethics Reader*. 2nd ed. London and New York: Routledge.

Ballantyne, R., Packer, J. and Hughes, K. (2009) Tourists' support for conservation messages and sustainable management practices in wildlife tourism experiences. *Tourism Management*. 30 (3), 658–664.

Balmford, A., Leader- Williams, N., Mace, G.M., Manica, A., Walter, O., West, C. and Zimmermann, A. (2007) Message received? Quantifying the impact of informal conservation education on adults visiting UK zoos. In A. Zimmerman (ed.) *Zoos in the 21st Century: Catalysts for Conservation*? Cambridge: Cambridge University Press, pp. 120–36.

Barney, E., Montez, J. and Yen, C. (2005) Assessing knowledge, attitudes and behaviour toward charismatic megafauna: the case of dolphins. *The Journal of Environmental Education* 36, 2, 41–55.

Beardsworth, A. and Bryman, A. (2001) The wild animal in late modernity: the case of the Disneyization of zoos. *Tourist Studies*. 1 (1), 83–104.

Bertella, G. (2014) The co-creation of the animal-based tourism experience. *Tourism Recreation Research*. 39 (1), 115–125.

Bethell, E. (2015) A 'how-to' guide for designing judgment bias studies to assess captive animal welfare. *Journal of Applied Animal Welfare Science*. 18 (1), S18–S42.

Born Free Foundation (2011). *The EU Zoo Inquiry Report Findings and Recommendations*. Available from http://endcap.eu/wp-content/uploads/2013/02/EU-Zoo-Inquiry-Report-Findings-and-Recommendations.pdf [Accessed 20 November 2017].

Breton, G. and Barrot, S. (2014) Influence of enclosure size on the distances covered and paced by captive tigers (Panthera tigris). *Applied Animal Behavioural Science*. 154, 66–75.

Bulbeck, C. (2005) *Facing the Wild: Eco Tourism, Conservation & Animal Encounters*. London: Earthscan.

Captive Animal Protection Society (2016) *Sad Eyes and Empty Lives* [online]. Available from: www.captiveanimals.org/wp-content/uploads/2011/02/Sad-Eyes-Empty-Lives.pdf [Accessed 11 April 2016].

Carr, N. (2016) Ideal animals and animal traits for zoos: General public perspectives. *Tourism Management*. 57, 37–44.

Christie, S. (2007) Zoo-based fundraising for in situ wildlife conservation. In A. Zimmerman (ed.) *Zoos in the 21st Century: Catalysts for Conservation?* Cambridge: Cambridge University Press, pp. 257–274.

Cohen, E. and Fennell, D. (2016) The elimination of Marius, the giraffe: humanitarian act or callous management decision? *Tourism Recreation Research*, doi.org/10.1080/02508281.2016.1147211

Cohn, J.P. (2006) Do elephants belong in zoos? *BioScience*. 56 (9): 714–717. doi: 10.1641/0006-

Croke, V. (1997) *The Modern Ark: The Story of Zoos, Past, Present and Future.* New York: Scribner.

Cunningham, E. (2016) (Re)creating forest natures: assemblage and political ecologies of eco-tourism in Japan's central highlands. In M. Mostafanezhad, R. Norum, E.J. Shelton and A. Thomson-Carr (eds) *Political Ecology of Tourism: Community, Power and the Environment.* Abingdon, Oxon: Routledge, pp. 169–187.

Dawkins, M.S. (2006) The Scientific basis for assessing suffering in animals. In P. Singer (ed.) *In Defence of Animals: The Second Wave.* Malden, MA: Blackwell Publishing, pp. 26–39.

De Koff, G. (1998) Conservation efforts of Australasian zoos – a review [online]. Unpublished report for Australian Regional Association of Zoological Parks and Aquaria (ARAZPA).

DeMello, M. (2014) The problem with zoos [online]. *The Dodo.* Available from: www.thedodo.com.community/mergobun/the-problem-with-zoos-603110345.html [Accessed 17 March 2016].

Dolmage, J. (2014) Is captive breeding actually helping the giant panda? [online]. *One Green Planet.* Available from: www.onegreenplanet.org/animalsandnature/is-captive-breeding-actually-helping-the-giant-panda/ [Accessed 4 February 2016].

Doward, J. (2016) Meat and dairy face the chop. *The Guardian.* 3 January. Available from: www.theguardian.com.

Duffy, R. and Moore, L. (2011) Global regulations and local practices: The politics and governance of animal welfare in tourism. *Journal of Sustainable Tourism.* 19 (4–5), 589–604.

Farm Animal Welfare Council (1979) The Five Freedoms. Available from: http://webarchive.nationalarchives.gov.uk/20121007104210/http:/www.fawc.org.uk/freedoms.htm [Accessed 8 August 2016].

Fennell, D.A. (2012) *Tourism and Animal Ethics.* London: Routledge.

Fennell, D.A. (2013) Tourism and animal welfare. *Tourism Recreation Research.* 38 (3), 325–340.

Ferdowsian, H.R., Durham, D.L., Kimwele, C., Kranendonk, G., Otali, E., Akugizibwe, T. and Johnson, C.M. (2011) Signs of mood and anxiety disorders in chimpanzees. *PLoS One*, 6, e19855. doi:10.1371/journal.pone.0019855

Fernandez, E.J., Tamborski, M.A., Pickens, S.R. and Timberlake, W. (2009) Animal-visitor interactions in the modern zoo: conflicts and interventions. *Applied Animal Behaviour Science.* 120 (1/2), 1–8.

Finch, J. (2008) Shoppers care more about animals than climate. *The Guardian.* 4 February. Available from: www.theguardian.com.

Franklin, A. (1999) *Animals and Modern Cultures: A Sociology of Human-Animal Relations in Modernity.* London: SAGE Publications.

Frost, W. (2011) *Zoos and Tourism: Conservation, Education, Entertainment?* Bristol: Channel View.

Hediger, H. (1964) *Wild Animals in Captivity.* New York: Dover.

Hughes, M., Newsome, D. and Macbeth, J. (2005) Case study: Visitor perceptions of captive wildlife tourism in a Western Australian natural setting. *Journal of Ecotourism.* 4 (2), 73–91.

Hutchins M, Smith B, and Keele, M. (2008) Zoos as responsible stewards of elephants. In C. Wemmer and C.A. Christen (eds) *Elephants and Ethics: Toward a Morality of Coexistence.* Baltimore: Johns Hopkins University Press, pp. 285–306.

International Union for the Conservation of Nature (IUCN) (2016) *Red List of Threatened Species.* Available from: www.iucnredlist.org/

Jamieson, D. (1995) Zoos revisited. In B. Norton, M. Hutchins, E. Stevens and T. Maple (eds) *Ethics on the Ark* Washington, DC: Smithsonian Institution Press, pp. 52–66.

Jamieson, D. (2006) Against zoos. In P. Singer (ed.) *In Defence of Animals: The Second Wave*. Malden, MA: Blackwell Publishing.

Keulartz, J. (2015a) Captivity for conservation? Zoos at a crossroads. *Journal of Agricultural and Environmental Ethics*. 28 (2), 335–351. doi:10.1007/s10806-015-9537-z

Keulartz, J. (2015b) Towards an animal ethics for the Anthropocene. In F.W Bovenkerk and J. Keulartz (eds) *Animal Ethics in the Age of Humans: Blurring Boundaries in Human-Animal Relationships*. Cham, Switzerland: Springer International Publishing, pp. 243–264.

King, R. 2005. The elephant whisperer. *Ecologist*. 35, 48–54.

Knight, J. (2009) Making wildlife viewable: habituation and attraction. *Society and Animals*. 17 (2), 167–184.

Konecki, K.T. (2008) Touching and gesture exchange as an element of emotional bond construction. Application of visual sociology in the research on interaction between humans and animals. Forum: *Qualitative Social Research*. 9 (3), Art. 33.

Kontogeorgopoulos, N. (2009) Wildlife tourism in semi-captive settings: A case of elephant camps in Northern Thailand. *Current Issues in Tourism*.12 (5–6), 429–449.

MacQueen, K. (2009) What Canadians really believe. *Maclean's*. 122 (6), 46–50.

Marino, L., Bradshaw, G. and Malamud, R. (2009) The captivity industry: the reality of zoos and aquariums. *Best Friends Magazine* [online]. 25–27.

Mason, G. (2006) Stereotypic behaviour: fundamentals and applications to animal welfare and beyond. In G. Mason and J. Rushen (eds) *Stereotypies in Captive Animals*, 2nd edition. Wallingford, UK: CAB International, pp. 325–356.

Mason, G. and Rushen, J. (eds) (2008) *Stereotypic Animal Behaviour: Fundamentals and Applications to Welfare*. Wallingford, UK: CAB International.

Masson, J.M., and McCarthy, S. (1995) *When Elephants Weep: The Emotional Lives of Animals*. New York: Delta.

Melfi, V. (2009) There are big gaps in our knowledge and thus approach to zoo animal welfare: a case for evidence-based zoo animal management. *Zoobiology*. 28 (6), 574–588.

Mellor, D. (2016) Updating animal welfare thinking: moving beyond the 'five freedoms' towards 'a life worth living'. *Animals*. 6 (3), 21.

Mintel (2014) *Visitor Attractions – UK December 2014* [online]. London: Mintel Group.

Moorhouse, T., Dahlsjo, C., Baker, S., D'Cruze N. and Macdonald, D. (2015) The customer isn't always right – conservation and animal welfare implications of the increasing demand for wildlife tourism. *PLoS ONE* [online]. 10 (10).

Neate, R. (2016) SeaWorld fights to restore its image as shares sink in the wake of Blackfish. *The Guardian*. 6 November. Available from: www.theguardian.com/environment.

Newsome, D. Dowling, R, and Moore, S. (2005) *Wildlife Tourism*. Clevedon: Channel View Publications.

Oluo, I. (2016) Cincinnati Zoo gorilla death: no child endangerment charges for mother. *The Guardian*. 6 June. Available from: Available from: www.theguardian.com.

Pekarik, A. (2004) Eye to eye with animals and ourselves. *Curator: The Museum Journal*. 47 (3), 257–260.

PETA (People for the Ethical Treatment of Animals) (2016) *Animals Used in Entertainment* [online]. PETA. Available from: www.peta.org/issues/animals-in-entertainment/animals-used-entertainment-factsheets/zoos-pitiful-prisons/ [Accessed 18 March 2016].

PFMA (Pet Food Manufacturers' Association) (2015) *Pet Population 2015* [online]. Available from: www.pfma.org.uk/pet-population-2015 [Accessed 31 March 2016].

Phillips, C.J.C. (2009) *The Welfare of Animals: The Silent Majority*. Dordrecht, The Netherlands: Springer.

Quinn, B. (2016) British tourist killed by elephant on trek in Thailand. *The Guardian*. 1 February. Available from: www.theguardian.com

Rabb, G.B. and Saunders, C.D. (2005) The future of zoos and aquariums: conservation and caring. *International Zoo Yearbook*. 39 (1), 1–26.

Robeck, T.R., Willis, K. Scarpuzzi, M.R. and O'Brien, J.K. (2015) Comparisons of life-history parameters between freeranging and captive killer whale (Orcinus orca) populations for application toward species management. *Journal of Mammalogy*. 96, 1055–1070.

Roe, K., McConney, A. and Mansfield, C. (2014) How do zoos 'talk' to their general visitors? Do visitors 'listen'? A mixed method investigation of the communication between modern zoos and their general visitors. *Australian Journal of Environmental Education*. 30 (2), 167–186.

Rothfels, N. (2002) *Representing Animals*. Bloomington, IN: Indiana University Press.

Shani, A. (2012) A quantitative investigation of tourists' ethical attitudes towards animal-based tourism. *Tourism*. 60 (2), 139–158.

Shani, A. and Pizam, A. (2008) Towards an ethical framework for animal-based attractions. *International Journal of Contemporary Hospitality Management*. 20 (6), 679–693.

Singer, P. (2016) Peter Singer on the animal rights revolution 4 decades after he started it. www.thedodo.com/peter-singer-on-the-animal-rig-726248280.html

Smith, L. and Broad, S. (2008) Comparing zoos and the media as conservation educations. *Visitor Studies*. 11 (1), 16–25.

Sommer, R. (2008) Semantic profiles of zoos and their animals. *Anthrozoos*. 21 (3), 237–244.

Therkelsen, A. and Lottrup, M. (2015) Being together at the zoo: zoo experiences among families with children. *Leisure Studies*. 34 (3), 354–371.

Thongma, W. and Guntor, B. (2011) Elephant camps and their impacts to community: Case study in Keud Chang, Chiang Mai Province, Thailand. *International Journal of Agricultural Travel and Tourism*. 2, 87–93.

Tribe, A. (2004) *Zoo Tourism. In Wildlife Tourism: Impacts, Management and Planning*, pp. 35–56. Melbourne: Common Ground.

Turesson, V. (2014) *On the Back of an Asian Elephant*. Student Report No. 587: Swedish University of Agricultural Sciences Department of Animal Environment and Health.

Turner, D. (2016) The Born Free Foundation. Available from: www.bornfree.org.uk (personal communication, April 2016).

Veltre, T. (1996) Menageries, metaphor and meanings. In R.J. Hoage and W.A. Deiss (eds) *New Worlds, New Animals: From Menagerie to Zoological Park in the Nineteenth Century*. Baltimore, MD: The Johns Hopkins Press, pp. 19–32.

Walters, J. (2015). Seaworld's attendance jumps despite latest cruelty allegations. *The Guardian*. 8 May. Available from: www.theguardian.com.

WAZA (World Association of Zoos and Aquariums) (2005) *Building a Future for Wildlife – the World Zoo and Aquarium Conservation Strategy*. Gland, Switzerland: WAZA.

WAZA (2016) www.waza.org/en/site/home [Accessed 8 August 2016].

Webster, J. (1995) *Animal Welfare: A Cool Eye Towards Eden*. London: John Wiley & Sons.

Woods, B. (2002) Good zoo / bad zoo: visitor experiences. *Anthrozoos*. 15 (4), 343–360.

World Animal Protection (2015). *Checking Out of Cruelty: How to End Wildlife Tourism's Holiday Horrors*. Available from: www.worldanimalprotection.org/sites/default/files/int_files/pdfs/checking_out_of_cruelty.pdf [Accessed 3 March 2016]

World Wildlife Fund (2016) *Endangered Species Directory* [online]. World Wildlife. Available from: www.worldwildlife.org/species/directory?direction=desc&sort=extinction_status [Accessed 1 March 2017].

YouGov UK (2012) *Animal Welfare Abroad* [online]. London: You Gov. Available from: https://yougov.co.uk/news/2012/08/07/animal-welfare-abroad [Accessed 25 October 2015].

Young, R.J. (2014) Zoos of the future break down the enclosure walls. *The Conversation*, 13 May. Available from https://theconversation.com/zoos-of-the-future-break-down-the-enclosure-walls-26605 [Accessed 1 March 2017].

5 Zoos and animal encounters

To touch or not to touch, that is the question

Neil Carr

Introduction

Zoos were originally constructed as sites of human entertainment. Indeed, Carr and Cohen (2011: 176) note that "The common theme throughout the histories of the zoo is of it as a place of human entertainment and leisure". It is against this background that touching and/or interacting with animals within zoos has come to be seen as a form of entertainment. This view has arguably been fed by zoos themselves who have in the past utilised images of people touching zoo animals as a means to promote the attractiveness of the zoo to tourists (see Figure 5.1 for example). However, the invitation to touch and interact with animals goes back further than promotional images like the ones from the 1970s displayed in Figure 5.1. Instead, visitors have been encouraged and facilitated in their desire to interact with zoo animals since the beginning of the contemporary zoos. These zoos have their roots in the late eighteenth and early nineteenth century when zoological societies such the Zoological Society of London were formed and subsequently opened the first truly public zoos (Turley 1998). These zoos offered

Figure 5.1 Promotional material for Durrell Wildlife Park from Jersey tourism brochures. From left to right, 1975 and 1976.

Source: left, Anonymous (1975), right, Anonymous (1976).

opportunities to ride elephants and camels and to witness events such as the chimpanzee tea party, perhaps the ultimate form of objectification of zoo animals for the entertainment of human visitors.

Since the use of the pictures in Figure 5.1 in the 1970s to market what was then Jersey zoo and has subsequently been renamed Durrell Wildlife Park, public opinion has shifted significantly regarding zoos. Today, there is widespread public distain for the idea that zoos should exist for no other reason than the entertainment of the public. Rather, zoos are now encouraged by social values to stress their role in the conservation of wild animals and their natural habitats (Reade & Waran 1996). This view is now commonly stated by those involved with zoos, including the World Association of Zoos and Aquariums (2005). In a continuous, self-reinforcing feedback loop, many zoos now stress their conservation work in their publicity material (Dibb 1995), encouraging the societal view of zoos as well as responding to it. The reason for the shift in the emphasis of zoos from entertainment to conservation relates to the rise of concern within society relating to questions of animal rights and welfare (Turley 1998). Rising concerns about the rights and welfare of animals are based on the notion that animals are sentient beings rather than unthinking and unfeeling objects (Bekoff 2007).

When animals were perceived to be little more than animated objects that lacked any sentience there was arguably nothing wrong in utilising them for entertainment. Conversely, once we appreciate animals as sentient beings we must recognise that they cannot simply be used for the entertainment of humans without any concern for their emotional and/or physical wellbeing. A welfarist perspective suggests that there is not necessarily anything inherently wrong with animals being involved in human entertainment, with the clear proviso that such entertainment should not be at the emotional and/or physical expense of the animal. Such entertainment should also not demean the animal. While the animal may not feel itself to be demeaned, humans, given their position of power within the human–animal relationship, are obligated to ensure the animal is not demeaned. Such a view has been espoused by Carr (2014) in relation to the treatment of dogs but the concept could easily be expanded to encapsulate all animals.

The hard animal rights perspective goes as far as to state that no animal should ever be used for/by humans but within an entertainment context and beyond. Indeed, Rudy (2011: 5) has stated that:

> The philosophers who advocate animal rights in the strong sense suggest that we should care for the animals we have here now, but that we stop breeding all domestic animals into existence for our use. No dogs, no cats, no horses, no pigs. No birds, no cows, no chickens, no fish. We wouldn't eat them or make them work for us or cuddle them or walk them or ride them or wear them. We couldn't use them to assist the blind or in search and rescue or to add joy to our lives.

Such a philosophical perspective gives primacy to the notion of the right of animals to live free from human subjugation. Consequently, the use of animals for human entertainment is wrong, from the hard animal rights perspective, irrespective of the impact of such activities on the emotional and/or physical wellbeing of the individual animal and potentially their entire species.

Possibly as a result of shifting social values, by the 1980s, as shown in Figure 5.2, images of humans touching animals were no longer employed to market Durrell Wildlife Park to potential visitors. While people were still featured in the promotional material they were now positioned at a distance from the animals, able to see them but not to touch them. This begins to purvey the notion that touching and interacting with zoo animals is not socially acceptable, unlike in the 1970s. This trend is similar, if more recent in its occurrence, to the changing presentations of bears in the natural environment in North America and the socially acceptable behaviour of visitors to these areas around the bears. Historically, just as in the case of zoo animals, the wild North American bears have been treated like and presented as objects to be utilised for the entertainment of humans. Indeed, up until

Figure 5.2 Tourism promotional material for Durrell Wildlife Park from Jersey tourism brochure (1986).

Source: Anonymous (1986).

the 1940s in Yellowstone National Park, viewing areas for visitors where bears were fed rubbish were provided and visitors feeding wild bears from their cars and out of their hands was a common occurrence well beyond then (Rugh 2008; Biel 2006). The latter was tacitly allowed and often actively promoted by the National Park agencies. This situation is illustrated in the words of Biel (2006: 1):

> The easy familiarity that developed between people and the park's [Yellowstone] most famous wild animals was addictive to visitors, and people fed the bears for many, many years. They fed them at garbage dumps, through car windows, and at campsites (purposely or not). They threw marshmallows. They offered cookies from the hand, sometimes coming away with a scar or two for their trouble. People fed the bears in spite of the fact that it was against the rules; they did it because it was 'fun.'

The shift away from treating the wild bear as an object for the entertainment of humans to a sentient being whose welfare should take precedence over the

Figure 5.3 Tourism promotional material for Durrell Wildlife Park (2008).
Source: Anonymous (2008).

desire of tourists for a close encounter has been driven by the same issues that have pushed zoos to change. This push has met resistance among tourists. Reports of people seeking to get close to bears in the wild and/or to feed them are still common in the twenty-first century. Indeed, during discussions with residents of Whistler, Canada, in 2016 it was noted that some tourists still try to feed the black bears in the region despite all the educational materials produced that explain why this is not appropriate behaviour.

The distancing of the zoo from the idea of humans not just touching but interacting with animals progressed, to reach its zenith in the early years of the twenty-first century when, as shown in Figure 5.3, adverts for Durrell Wildlife Park were bereft of human visitors. Here no expectation is raised at all of touching an animal when visiting the zoo. An underlying message is that these animals are not merely objects for the titillation of people during their leisure time.

In line with the shift away from defining themselves as sites of entertainment where it was acceptable to interact with the animals, when visiting zoos today it is common to see signs extolling people not to touch, feed, or interact with the animals in any way, as seen in Figure 5.4. However, the signs depicted in Figure 5.4 can also be interpreted as admittance by zoos that visitors still seek to touch and interact with animals; even in the face of negative views regarding doing so and the remodelling of zoos away from their previous perceived and often real focus on human entertainment.

Figure 5.4 No interacting with the animals (from top left to bottom right: Singapore Zoo, Alpenzoo Innsbruck, ZooParc de Beauval France, Edinburgh Zoo). (All signs photographed in 2013.)

An integral component of the image of zoos as centres of conservation is how they have been constructed as places capable of educating the public about the importance and value of conservation work (Serrell 1981; World Association of Zoos and Aquariums 2005; Ballantyne *et al.* 2007; Smith and Broad 2008). The ability of zoos to offer an effective learning experience relates to their ability to provide informal and unstructured learning opportunities, which in turn allow for free-choice learning (Falk 2005). These, it is argued, are more effective learning tools than traditional structured learning in schools (Reade & Waran 1996). Although it may stand in apparent conflict to the images in Figure 5.4 it has been suggested that zoos have the best potential to educate the public when they offer some form of physical and/or emotional interaction between the learner and animals (Clayton *et al.* 2009; Lindemann-Matthies & Kamer 2006; Swanagan 2000; World Association of Zoos and Aquariums 2005). It is within this context that Morgan and Tromborg (2007: 280) have stated that "To be fair to zoos, the underlying intention for increased visitor–animal interaction is to promote the empathy and sense of connectedness needed to improve attitudes and behavior with respect to conservation". In comparison to these active learning opportunities, passive educational experiences offered via simply viewing animals from a distance or reading notice boards are less effective methods for engaging people in conservation messages (Carr & Cohen 2011).

While many zoos have sought to reposition themselves as centres of conservation they are arguably still driven by the reality that they are dependent on the money they generate from visitors (Dibb 1995; Davey 2007; Hosey 2008). Furthermore, while the specific nature of the entertainment demanded by visitors may have altered, the primary reason for visiting zoos is still entertainment (Fernandez, *et al.* 2009; Puan & Zakaria 2007; Sickler & Fraser 2009). Consequently, entertainment, albeit in a potentially driven guise to that offered historically, remains fundamentally important to the economic viability of zoos.

Furthermore, for zoos to act as effective sites of conservation education it has been suggested that visitors must be entertained (Jamieson 1985). Within this context, Puan and Zakaria (2007, p. 232) have stated that "It is undeniable that effective education can only be gained if the desire of visitors for enjoyment is met". Yet such a view is arguably missing the point that in order for learning opportunities to be effective, inside or outside of the classroom, they need to be entertaining in that they engage the learner and make them into a willing and active learner. In effect this is the opposite of what occurs in a boring, unengaging learning experience. Perhaps the ultimate example of such an experience is provided by the history of magic teacher, Professor Binns, at Hogwarts in the Harry Potter series. The professor manages to put students to sleep via his constant monotonous monologue rather than instilling in them an interest in bloody goblin rebellions. In such a learning experience there is a lack of engagement with and by the supposed recipients of the learning. Such an experience becomes unentertaining. Indeed, Sickler and Fraser (2009) go further by suggesting that learning can be fun/entertaining. This combination has been referred to as 'edu-tainment' and is employed in a variety of tourism settings, including museums and zoos (Carr 2011). It is within this context that the World Association of Zoos and Aquariums (2005) has suggested that mixing conservation education and entertainment is not in itself problematic, a view supported by Ballantyne *et al.* (2007). The task for zoos then is how to ensure the right balance is struck between entertainment and education such that they complement each other rather than one dominating the other. If the balance swings too far to entertainment then any educational message can be lost. On the other hand, if the balance swings too far to traditional, boring, teaching methods then the lessons may fail to be learned.

In relation to touching animals and interacting with them, zoos have arguably aimed to strike the balance between entertainment and education by offering animal interaction and touching experiences within controlled environments and under the guidance and observance of educational and zoological experts. Opportunities to still touch animals are not entirely uncommon within zoos and can include reptile encounters where an opportunity is offered to zoo visitors to learn about and gently touch something like a corn snake. Whether such an experience meets the desires of people to experience the animals they are most attracted to (see Carr 2016a and 2016b for details) is another question. Zoos are also increasingly catering to the desire of visitors to get up close to animals by offering them 'behind the scenes' and 'keeper for a day' experiences. While these may offer the opportunity to get close to those animals visitors see as attractive, questions can be raised about the balance between entertainment and education involved in such experiences, something that clearly needs studying. An additional potential problem with behind the scenes experiences and animal encounters overseen by zoo staff is that they may place in the minds of visitors an image of the zoo that contradicts the message about not touching or interacting with the animals. In this context there is the question of why a visitor should not be able to touch or get up close to an animal whenever they want when they see such opportunities being offered in the zoo. The distinction may be said to be that one is under the control of the authority of the zoo staff while the other is not, but such a distinction may be missed potentially though not necessarily deliberately, by visitors eager to experience a close encounter with a wild animal. The same problem can be said to be related to the feeding of zoo animals by staff while in the presence of visitors. Such practices can arguably raise the question in the mind of visitors of why they cannot feed the animals if other people (i.e., staff) can.

Overall, what we see is that while, historically, attempts to interact with animals in zoos and/or touch them were often allowed and encouraged by many zoos, today such behaviour is constructed as deviant. This past is painted as being associated with the abuse of animals for the entertainment of human visitors. However, touching and interacting with animals in zoos has not entirely ceased. Rather, when facilitated by zoos today such opportunities are constructed as an educational experience: one that is fun/entertaining, but only for the purpose of aiding learning. At the same time, at least some of the animal interaction experiences offered by many zoos today are clearly generating income above and beyond that generated by people simply passing through the entrance to the zoo (Kreger & Mench 1995). It is against this background that this chapter seeks to explore the extent to which visitors to zoos still attempt to interact with animals outside of structured and controlled opportunities provided by zoos. The chapter also looks at potential explanations for why visitors to zoos ignore the type of signs highlighted in Figure 5.4 and attempt to interact with zoo animals.

Methods

The data on which this chapter is based was gathered using a non-participant observation method during a study of the attractiveness of different types of animals to visitors to Durrell Wildlife Park. The park is located on the island of Jersey, off the south coast of England. It was opened as Jersey Zoo in 1958 by the late Gerald Durrell and today attracts approximately 180,000 visitors each year (Anonymous 2015). The Park is spread over approximately 30 acres of land and is home to an array of animals, the largest of which are the gorillas and orangutans (more details about the Park can be found at http://www.durrell.org/visit/).

Details of the animal attractiveness study, and the resultant conclusions, can be found in Carr (2016a). Although the study was focused on assessing how attractive different types of animals were to visitors, instances of visitors attempting to interact with animals were also recorded. The study was conducted over a period of two and a half weeks during the peak summer vacation period between late July and mid-August 2013. Observation was undertaken at specific animal enclosures across Durrell Wildlife Park. Each enclosure was studied during three separate 90–minute periods. The first was between the opening of the Park and the beginning of the lunch period ('morning'), the second included the lunch period ('lunch'), while the third period covered the afternoon and ceased one hour before the closure of the park ('afternoon'). The precise timing of the observations altered slightly in order to avoid any observations being gathered when an animal talk by the educational officers at Durrell was taking place.

Box 5.1 identifies the locations of the observations and the animals involved. The animals were divided according to their taxonomic classification to ensure a range of different animal types were studied. Where multiple animals are listed within a single observation period, this is because they inhabit the same enclosure. An example of this is the orangutans and gibbons.

Box 5.1 Animals focused on during non-participant observation

- Orangutans and gibbons at information hut
- Orangutans and gibbons at inside area
- Meerkats
- Otters
- Fruit bats

(Continued)

(Continued)

- Gorillas
- Burmese pythons and green crested lizards
- Galapagos tortoises
- Meller's ducks and Bali starlings
- Red river hogs

Given the nature and size of many of the enclosures and the viewing areas offered around them it was impossible in many instances for the researcher to observe all visitors who may have been looking at a particular animal at any one time. Consequently, the data is more precisely related to the area under observation than the entire area surrounding the enclosure of the animals. Given the limited ability to observe the whole area surrounding many of the enclosures the researcher was positioned at a specific location in relation to the animals and did not move from this area throughout the periods of observation. The decision to have two study sites associated with the orangutans and gibbons was related to the fact that their enclosure is one of the most diverse (both for the animals and the viewing public).

During the observation periods an individual or group was identified as they entered the research area and observed remotely until they exited the area. A convenience sampling method was employed to identify people to be observed. The 30 observation sessions undertaken resulted in 498 separate observations. In addition to these observations a series of incidental observations was also undertaken during the same periods. These incidental observations were initially unplanned and only triggered by the researcher noticing behaviour of visitors outside of those identified for observation on entering the research area that it was felt would be interesting to record. After these incidental observations began, while the focus remained on the sampling method noted above, they continued intermittently throughout the data collection exercise. The visitors involved in these incidental observations were not consistently monitored during their time in the research area. No information was gathered about the frequency with which incidental observations took place as they were not the focus of the study but instead may be best described as spontaneous or opportunistic moments of data collection. It should be noted that this research had the appropriate ethical clearance.

Results

Table 5.1 identifies the number of visitors observed within the research areas who attempted to gain the attention of, interact with, or feed the animals. This data shows that very few of the people observed within the research areas either attempted to gain the attention of the animals or interact with them. Similarly, very few of the visitors were observed feeding the animals. Consequently, it may be suggested based on the data presented in Table 5.1 that hardly any of the

Table 5.1 Visitor attempts to interact with animals

Animals/research areas[a]	*Visitors attempting to interact with animals (%)*	*Visitors attempting to gain animal attention (%)*	*Visitors feeding animals (%)*
Orangutan (information hut) ($n = 49$)	1 (6.25)	(0.00)	0 (0.00)
Orangutan and gibbon ($n = 30$)	4 (13.33)	(0.00)	0 (0.00)
Meerkat ($n = 64$)	0 (0.00)	7 (10.94)	0 (0.00)
Otter ($n = 34$)	0 (0.00)	(0.00)	0 (0.00)
Flying fox/fruit bat ($n = 37$)	0 (0.00)	(0.00)	0 (0.00)
Gorilla ($n = 51$)	0 (0.00)	(0.00)	0 (0.00)
Burmese python and green crested lizard ($n = 71$)	0 (0.00)	(0.00)	0 (0.00)
Galapagos tortoise ($n = 56$)	4 (7.14)	4 (7.14)	2 (3.57)
Meller's duck and Bali starling ($n = 55$)	0 (0.00)	(0.00)	(0.00)
Red river hog ($n = 51$)	0 (0.00)	2 (3.92)	(0.00)

Note

a n = total number of observations.

visitors to Durrell Wildlife Park attempt to engage in the deviant behaviour of attempting to interact with, gain the attention of, or feed the animals. This may be portrayed as a positive result for the visitor education signs noted in Figure 5.4 and for the attempts to change the image of zoos from places of entertainment via animal interaction/contact to ones of conservation and education discussed earlier in the chapter.

The results noted in Table 5.1 were supplemented by data collected via the incidental observations. The observation notes collected during the incidental observations clearly indicate that some other visitors either attempted to or actually managed to interact with, gain the attention of, or feed the animals, indicating that frequency of this behaviour is higher than depicted by Table 5.1. Yet even with these additional people it is clear that engaging in any or all of these deviant behaviours is a minority behaviour. However, while engaged in by only a minority, such behaviours are still potentially dangerous to the health and wellbeing of the animals, as well as the zoo visitors.

It is also interesting to note that while interacting with, feeding, and/or touching zoo animals has been defined as deviant behaviour, none of the visitors who did not engage in this form of behaviour and conformed to the rules of the zoo appeared to seek to prevent anyone from engaging in such deviant behaviour. Indeed, the only instance noted where a visitor was prevented from continuing in their behaviour of attempting to touch the tortoises was the result of a zoo employee noticing their behaviour.[1]

Tapping on the glass of the enclosures of the various animals located in the reptile house and the glass viewing dome in the meerkat enclosure were activities that were observed on several occasions and represent the most common attempts to gain the attention of animals. This behaviour was undertaken by both adults and children. Attempts to interact with the animals were generally more remote, entailing the visitors waving at animals (particularly the orangutans). The one exception was the significant number of people, both adults and children, who attempted to touch the tortoises. Attempts to gain the attention of animals appear to have been made for a variety of reasons, including making the animal active, entertaining the children and gaining the perfect picture.

Potentially most problematic of the observed behaviours were the attempts made to feed the animals. This behaviour, as highlighted in Table 5.1, was almost exclusively associated with the tortoises. The only other animal enclosure where any attempt at feeding was witnessed was that of the Red River hogs. In the case of both animals, visitors attempted to give them vegetation that was immediately outside of the enclosure.

The number of people touching and feeding the tortoises may appear to stand in opposition to the popularity of these animals (see Carr 2016a). However, an examination of the structure of Durell Wildlife Park and the enclosures within it shows it may not be the popularity of an animal that determines the number of visitors who seek to touch and feed them but how accessible they are. The feeding and touching of the tortoises was clearly a relatively easy activity for the visitors, with the tortoises content to come very close to the fence and the fence itself being an insignificant barrier to prevent the tourists touching and feeding the tortoises. In contrast, the more popular animals, such as the gorillas and orangutans (Carr 2016a) are very difficult to touch or feed. This does not, however, stop visitors from trying to get as close as possible to all the animals. For example, it was very common at the gorilla enclosure to see adults leaning over the wall surrounding the enclosure and not rare to see children climbing the wall or parents lifting them onto the top of the wall to gain a better view of the animals despite the real danger of them potentially falling into the enclosure.[2] Similar behaviour was witnessed at the otter enclosure (see Figure 5.5) where the fence seemed an irresistible climbing wall for many children and for most adults was something to lean over to gain a better view of the animals.

It is clear from the observations undertaken if a fence is made of a substance that little fingers can hold onto then that is exactly what young children will do. There were hardly any young children who visited the tortoise enclosure who did not curl their fingers through the mesh fence. These children were not placing their fingers in this position to attract the tortoises or apparently giving any conscious thought to this behaviour. Rather, it seems they were simply holding onto the fence. The problem is that those fingers when curled around the fence are at the right height for inquisitive tortoises to reach with their mouths.

The desires of zoo visitors, as reported by Carr (2016a, 2016b, 2016c), are clearly to be able to see the animals, the closer the better, and for the animals to be active. Results from the non-participant observations detailed in this chapter clearly indicate that if the opportunity is there, then a portion of visitors are willing to take advantage to get as close as possible to the animals, gain the 'ideal' picture in the process and see the animals being active. This is exemplified in

Figure 5.5 Interactions with the otters.

Figure 5.6 where visitors were observed stepping off the path and over a small fence to get as close as possible to the meerkat enclosure.

Discussion and conclusions

The desire amongst visitors to interact with and get close to animals, potentially endangering their own health and that of the animals, was not a focus of the study from which this chapter has developed. Rather, it was an emergent finding; one that arguably needs further investigation. The material presented in this chapter suggests there is a component of the visitor population to zoos, because I see nothing to suggest that Durrell Wildlife Park is unique in this respect, which may need more careful managing to prevent serious injury to animals and/or visitors. It could be argued that better fencing would prevent the access to the tortoises that visitors are currently able to exploit at Durrell Wildlife Park. Similarly, more fencing may prevent scenes like those noted in Figures 5.5 and 5.6. However,

Figure 5.6 Interactions with the meerkats.

erecting more and bigger fences flies in the face of the modern image of the zoo as something other than a prison for animals. This is despite the fact that the fences in this instance may be an option for the safety of the animals rather than the safety of the visitors (i.e., keeping the fearsome wild beast caged and therefore the human visitor safe, or keeping the fearsome beast caged and therefore the animals safe). However, erecting fences does not get at the underlying problem, that visitors want to engage in this behaviour and either see nothing wrong in doing it or feel they have the right to as humans and/or paying visitors to the zoo. These issues speak to the humanist view of the world that sees us as the dominant creature on the planet and that everything (including the animals) on it are there for us.

The continuation of the existence of such humanist views can appear to be contrary to suggestions that society is increasingly aware of the rights and welfare needs of non-human animals as sentient beings. Alternatively, they may be presented as evidence of the fact that changes in social and cultural values are always a 'work in progress'. Consequently, old views are not simply consigned to the dustbin of history when new ones emerge. Instead, they tend to linger, overlapping with newly emergent ideas. In other words, the desire to be entertained by animals in zoos can sit side by side, at both the individual and societal level, with an appreciation of animal rights and welfare that defines such desires as deviant.

If the only reason we continue to see a minority of people seeking to engage with animals in zoos is the lingering of old, previously socially acceptable, human values then arguably all we need to do is wait a little longer to see these values die and be replaced with the new values that define such behaviour as deviant. However, at least in the case of zoos such an approach is overly simplistic. It ignores the fact that under the guise of education and under the supervision of zoo staff, touching and interacting with animals is both allowed and encouraged in the contemporary era. This arguably sends out a mixed message that interacting with animals in zoos is both allowed and not allowed. The distinction may be that it is allowed under expert supervision but while technically true the images portrayed in the two types of behaviour are fundamentally the same. The consequent argument is that animal interaction under the supervision of experts may actually help to prolong the notion from a previous era of the zoo as a place where it is socially acceptable to interact with the animals, with or without expert supervision.

So, should all contact between animals and visitors in zoos be banned? Two problems raise their heads in response to this suggestion. The first is that the contemporary experiences offered by zoos where visitors are able to interact with animals offer a financial income for zoos that are dependent on satisfying visitors for their financial viability. Secondly, many of the opportunities zoo visitors have now to interact with animals in a controlled manner have been constructed and presented to society in an educational context. Consequently, they are an integral part of the rebranding of zoos discussed earlier in the chapter in response to changing social values regarding animals. Yet these educational offerings are not merely a marketing ploy. They are built on the concept that such experiences can lead to a more successful learning experience, with the result that zoos truly can help save animals and their natural habitats from destruction. So the problem we are left with is that interacting with animals may be a way to help preserve their species and natural environment. At the same time, interacting with animals can cause animals to suffer. Furthermore, controlled and monitored interaction with animals may actually help to support traditional views that interacting with wild animals for human enjoyment is socially acceptable. More research is clearly needed to explore the validity of this suggestion and to attempt to find solutions. What is clear at the moment is that uncontrolled interactions with animals within zoos continue to take place. Efforts to stop these interactions need to continue, but not at the expense of depriving zoos of one of their best tools for raising visitor awareness of the need to care for the wild brethren of the zoo animals and their natural environments.

The implications of touching and not touching zoo animals for the welfare and rights of the individual animals and their species are multiple. Inappropriate contact/interaction between visitors and zoo animals is clearly stressful for the latter, which has implications for their wellbeing (Fernandez *et al.* 2009; Hosey 2000). However, the potential exists for controlled animal encounters to be non-stressful to animals and potentially even enriching to some. Beyond these individual animals if interacting with an animal can benefit the future survival of their wild brethren through education of visitors about the importance of wildlife and natural environment conservation then this must be assessed under the purview of animal welfare. However, such wide-ranging welfare should not be at the expense of the individual animal. Balancing these issues is clearly a challenge for zoos if they are to overcome their critics and have a future beyond entertaining humans in the twenty-first century.

Acknowledgement

I would like to thank Durrell Wildlife Park for helping to facilitate the research on which this chapter is based.

Notes

1 There was an ethical dilemma regarding the attempts by some visitors to engage with the animals. Should any attempts have been made by the researcher to stop these behaviours? After consulting several members of the Durrell staff the decision was made not to do so, on the basis that the researcher was not a Durrell employee. While the logic of this position is sound and not intervening allowed the behaviour to be observed it is still a decision that I am not entirely comfortable with.

2 On several occasions zoo staff were observed ‘suggesting’ to parents that it was not best for their children to be positioned on the top of the wall surrounding the gorillas.

References

Anonymous (2015). Visitor numbers falling at Durrell – could it be forced to close. *Jersey Evening Post*. http://jerseyeveningpost.com/news/2015/02/13/visitor-numbers-falling-at-durrell-could-it-be-forced-to-close/ [accessed 15.7.15].

Anonymous (2008). *Destination Jersey*. Available at Jersey Archive. Ref: D/AS/E8/12.

Anonymous (1986). *What's on Jersey*. Available at Jersey Archive. Ref: D/AS/A4/15.

Anonymous (1976). *Jersey*. Available at Jersey Archive. Ref: D/AS/A4/23.

Anonymous (1975). *Jersey*. Available at Jersey Archive. Ref: D/AS/A9/1.

Ballantyne, R., Packer, J., Hughes, K. & Dierking, L. (2007). Conservation learning in wildlife tourism settings: lessons from research in zoos and aquariums. *Environmental Education Research*. 13(3): 367–383.

Bekoff, M. (2007). *Animals Matter*. Shambhala. Boston.

Biel, A. W. (2006). *Do (Not)Feed the Bears: The Fitful History of Wildlife and Tourists in Yellowstone*. University Press of Kansas. Lawrence, KA.

Carr, N. (2016a). Star attractions and damp squibs at the zoo: A study of visitor attention and animal attractiveness. *Tourism Recreation Research*. 41 (3): 326–338.

Carr, N. (2016b). An analysis of zoo visitors' favourite and least favourite animals. *Tourism Management Perspectives*. 20: 70–76.

Carr, N. (2016c). Ideal animals and animal traits for zoos: General public perspectives. *Tourism Management*. 57: 37–44.

Carr, N. (2014). *Dogs in the Leisure Experience*. CABI. Wallingford, UK.

Carr, N. (2011). *Children's and Families' Holiday Experiences*. Routledge. London.

Carr, N. & Cohen, S. (2011). The public face of zoos: Balancing entertainment, education, and conservation. *Anthrozoos*. 24 (2): 175–189.

Clayton, S., Fraser, J. & Saunders, C. (2009). Zoo experiences: Conversations, connections, and concern for animals. *Zoo Biology*. 28: 377–397.

Davey, G. (2007). An analysis of country, socio-economic and time factors on worldwide zoo attendance during a 40 year period. *International Zoo Yearbook*. 41: 217–225.

Dibb, S. (1995). Understanding the level of marketing activity in the leisure sector. *The Service Industries Journal*. 15(3): 257–275.

Falk, J. (2005). Free-choice environmental learning: Framing the discussion. *Environmental Education Research*. 11(3): 265–280.

Fernandez, E., Tamborski, M., Pickens, S. & Timberlake, W. (2009). Animal-visitor interactions in the modern zoo: Conflicts and interventions. *Applied Animal Behaviour Science*. 120: 1–8.

Hosey, G. (2008). A preliminary model of human–animal relationships in the zoo. *Applied Animal Behaviour Science*. 109: 105–127.

Hosey, G. (2000). Zoo animals and their human audiences: What is the visitor effect? *Animal Welfare*. 9: 343–357.

Jamieson, D. (1985). Against zoos. P. Singer (ed.) *In Defence of Animals*. Basil Blackwell. Oxford, pp. 108–117.

Kreger, M. & Mench, J. 1995. Visitor-animal interactions at the zoo. *Anthrozoos*. 8 (3): 143–158.

Lindemann-Matthies, P. & Kamer, T. (2006). The influence of an interactive educational approach on visitors' learning in a Swiss zoo. *Science Education*. 90 (2): 296–315.

Morgan, K. & Tromborg, C. (2007). Sources of stress in captivity. *Applied Animal Behaviour Science*. 102: 262–302.

Puan, C. & Zakaria, M. (2007). Perception of visitors towards the role of zoos: A Malaysian perspective. *International Zoo Yearbook*. 41: 226–232.

Reade, L. & Waran, N. (1996). The modem zoo: How do people perceive zoo animals? *Applied Animal Behaviour Science*. 47: 109–118.

Rudy, K. (2011). *Loving Animals: Toward a New Animal Advocacy*. University of Minnesota Press: Minneapolis.

Rugh, S. (2008). *Are We There Yet? The Golden Age of American Family Vacations*. University Press of Kansas. Lawrence, KA.

Serrell, B. (1981). The role of zoological parks and aquariums in environmental education. *Journal of Environmental Education*. 12 (3): 41–42.

Sickler, J. & Fraser, J. (2009). Enjoyment in zoos. *Leisure Studies*. 28 (3): 313–331.

Smith, L. & Broad, S. (2008). Do zoo visitors attend to conservation messages? A case study of an elephant exhibit. *Tourism Review International*. 11: 225–235.

Swanagan, J. (2000). Factors influencing zoo visitors' conservation attitudes and behavior. *The Journal of Environmental Education*. 31(4): 26–31.

Turley, S. (1998). Exploring the future of the traditional UK zoo. *Journal of Vacation Marketing*. 5(4): 340–355.

World Association of Zoos and Aquariums. (2005). *Building a Future for Wildlife: The World Zoo and Aquarium Conservation Strategy*. WAZA Executive Office. Bern.

6 Being Camilla

The ‘leisure’ life of a captive chameleon

Samantha Wilkinson

Introduction

According to the Shona and Matabele tribes of northern and southern Zimbabwe, respectively, the Great Spirit in the Sky sent a rabbit to inform the tribes about the location of herds of buck, which they could hunt in order to feed their women and children (Alexander, 2002). Yet, the rabbit ran too quickly, only managing to reach one village before experiencing tiredness and hunger. Consequently, the rabbit sent a chameleon to the remaining villages. Yet, by the time the villagers received the message, and reached the place where the herds allegedly were, the animals had moved on. Consequently, today, the Shona and Matabele tribes believe that if someone is touched by a chameleon, they will have bad luck, and be hungry, for the remainder of their life (Alexander, 2002). This is a story about animals ‘doing work’, or labour’, intersecting with human labour and leisure interests. In the absence of human effort, the perceived poor undertaking of an animal labour task, on behalf of deities and humans, has led to stigmatisation of chameleons in folklore.

This chapter is based on research I conducted into the ‘leisure’ life of Camilla, a captive veiled chameleon, who resides in the ‘Creepy Critters’ hut at Vision Kingdom[1] Theme Park. In this chapter, I aim to enhance understanding of human–reptile behaviour within the leisure experience. In order to do so, I draw on two innovative methods: autoethnography, and video-elicitation focus groups. Through the approach of egomorphism – in which I draw on personal experiences to read Camilla – I offer insight into the subjectivities and everyday life of Camilla, with a focus on her feelings during her ‘leisure’ time. For instance, when she is happy, seeks attention, or wishes to be left alone. Further, through video-elicitation focus groups with visitors to the Park, I find that Camilla is affect*ive*; she has the ability to engender diverse emotions in people during their leisure time, including: relaxation, fascination and pity. Additionally, participant observation shows that Camilla is affect*ed*; humans stimulate emotions in Camilla, including fear and relaxation.

This chapter is underpinned by ongoing debates about the welfare of captive creatures in leisure spaces. I question whether Camilla can be said to have ‘leisure’ time when her spatial freedom is restricted, and when under the constant gaze of leisure users. By considering Camilla as an affective, sentient creature, with the capacity to influence human emotions, this chapter offers a departure from much existing work which tend to consider animals as passive within the leisure experience (Carr, 2010). The term 'leisure' is subject to multiple and often competing definitions, yet can be broadly defined as "free or unobligated time that does not involve work or performing other life sustaining functions" (Leitner and Leitner, 2012, 3). This chapter concludes that human visitors are at leisure, and Camilla is a worker facilitating this leisure experience for visitors. I will now detail Camilla's personal situation, before going on to review relevant literature.

Camilla's personal situation

First, I wish to bring attention to the fact that Camilla is a named reptile. According to Levin (2015), the act of naming conveys quasi-personhood; it suggests that the reptile is unique and cannot be replaced (Levin, 2015). Camilla, a 15-month-old female veiled chameleon (see Figure 6.1), resides in the 'Creepy Critters' reptile house at Vision Kingdom Theme Park. Veiled chameleons (*Chamaeleo calyptrarus*) are an arboreal lizard species that range from Asir Province, southwestern Saudi Arabia, to Aden, Yemen, where they live on high, dry plateaus (Schmidt, 2001). A distinguishing feature of veiled chameleons is their ability to change colour in response to fluctuations in light, shade and temperature, and as an expression of emotion (Mattison and Garbutt, 2012). Veiled chameleons use their opposing toes and strong tails to help them cling to branches (Berre, 2009), and their protruding sticky tongue to target prey in one-fiftieth of a second (Stewart, 2008). Further, veiled chameleons' eyes often move independently, enabling them to observe two different objects simultaneously (Krysko *et al.*, 2004). As Bartlett and Bartlett (2001) note, veiled chameleons are solitary reptiles, pairs are often aggressive to one another, and thus Camilla lives alone.

Camilla was bought from a local chameleon breeder, by the Theme Park owners, at the age of three months, and has since resided in the Creepy Critters hut (see Figure 6.2). The naming of this hut is not without significance, serving to promote Camilla, and the other inhabitants (see Figure 6.3), including a corn snake, Chinese water dragon, Burmese python, bearded dragon and spectacled caiman, as 'odd' entities.

Upon walking into the Creepy Critters hut, one is greeted by an atmospheric tropical rainforest soundtrack. While such environmental manipulation is intended to create a simulacrum of the wild (Neo and Ngiam, 2014), devices such as darkening the viewing area, whilst lighting up each exhibit, are also for the benefit of leisure users, in that it provides both optimum visibility and heightened dramatic effect (cf. Markwell and Cushing, 2009). Inside Camilla's 4 ft by 4 ft glass vivarium, simulation of the natural habitat chameleons live in is also achieved through the creative use of artificial plants and branches (see Figure 6.4). Further, as Figure 6.5 shows, each vivarium is accompanied by a brief explanation of its contents.

Figure 6.1 Camilla.

Figure 6.2 Creepy Critters hut.

Figure 6.3 Camilla's vivarium (centre) and surrounding vivaria.

Figure 6.4 Camilla in her vivarium.

Figure 6.5 Information board below Camilla's vivarium.

Figure 6.6 Camilla inflates her body in a territorial display.

Figure 6.7 Camilla preparing to target prey.

Figure 6.8 Camilla's projectile tongue snapping up prey.

Figure 6.9 Camilla eating brown cricket.

It is noteworthy that no information is provided on the specific individual in these fact files. Visitors are only likely to learn Camilla's name and gender by attending 'meet the creepy critters', an activity occuring at 11.30 a.m. daily, during which the staff who care for the reptiles provide further details on them.

Having detailed Camilla's personal situation, I now outline literature on 'individual animals', inter-species affective encounters, and egomorphism, respectively. I am engaging with this material because the experiences of individual animals in leisure experiences has been somewhat sidelined, in favour of collectivities, such as species and herds (Bear, 2011). Further, by exploring inter-species affective encounters, I demonstrate that chameleons are far from passive in human leisure experiences. Finally, egomorphism is an approach which uses the *self* as the primary reference point for understanding animals (Milton, 2005); in comparison to anthropomorphism, which is based on the attribution of *human* characteristics to animals. As leisure activities are often self/ego focused, egomorphism may be an appropriate lens through which to gain insight into human–animal leisure experiences.

Individual animals in leisure spaces

Scholars adopting an animal geography perspective are now legion. However, the focus on collectivities elides differences between individual animals (Bear, 2011). When researchers have focused on individual animals, it has typically been furry pets, of which Goode's (2006) account of play with his dog Katie is a fine example. There are, however, a few notable exceptions. For instance, Whatmore (2002, 231), through an ethnographic account of Duchess and Gay, elephants at Paignton Zoo, illustrates that these animals are "active agents who make a difference to the ways in which … heterogeneous social networks take and hold their space". Whatmore (2002, 5) investigated different "becomings" of elephant, with a focus on Duchess's experiences of moving to a new enclosure that aimed to mimic her home habitat of the African savannah. However, Philo (2005, 829) critiques Whatmore's (2002) writings on elephants, contending that the animals "still remain somewhat shadowy presences". Whatmore's (2002) findings could potentially have thus been of greater interest if she had conducted more in-depth observations of the individual animal.

More recently, Levin (2015) wrote a paper on Marius the giraffe. The author explored the decision of Copenhagen Zoo to kill Marius in the name of population management. Levin (2015) argues that the zoo adopted a variety of display practices in the first two years of Marius's life, treating him as 'special'. Here, the word 'special' involves displaying the animal as an individual; having a unique backstory of how he got to the zoo; a unique personality and set of physical dispositions (Levin, 2015). The author contrasts this with the justification for killing Marius, which referenced zoo objectives, and treated Marius as a mere specimen; in other words, a member typical of a class. Studies of individual mammals in the animal geographies literature are more prevalent than studies of non-mammals. Recognising this, Bear (2011) offered an ethnographic account of a giant Pacific, captive, octopus in The Deep aquarium. Of particular importance here is Bear's (2011, 298) attempt to draw attention to this non-mammal's individuality, stating:

> Being a giant Pacific octopus is much like being a 'human' – a number of beings with similar characteristics, but that are nonetheless significantly different from each other. Ostensibly, the octopus is here as a representative of its species. However, this is not just a giant Pacific octopus; this is Angelica.

Despite praising Bear (2011) for paying attention to a non-mammalian life form in leisure spaces, the author could have perhaps made more of the contingency of all actants involved in the leisure space (for instance, staff, visitors, other sea life and the tanks), and stressed their relationality more. By doing so, he may have made the politics of confinement a little more explicit. One means of engaging with the relationality of actants in leisure experiences is to pay attention to inter-species affective encounters.

Inter-species affective leisure encounters

Humans engage with animals through a range of affective exchanges (Jones, 2011). Yet, as Thrift (2004) articulates, 'affect' has no stable definition. In a field of literature whereby 'affect' is difficult to comprehend, Whatmore's (2006, 604) definition is a useful one. 'Affect' refers to "the force of intensive relationality – intensities that are felt but not personal; visceral but not confined to an individuated body". This understanding of affect can be both tied to, and distinguished from, feeling. Affects occur between bodies, and these affects find "corporeal expression in bodily feelings" (Anderson, 2006, 736). Following this line of thought, 'affect' can be seen as "a kind of vague but intense atmosphere" and 'feeling' as "that atmosphere felt in the body" (McCormack, 2008, 6). Further, both 'affect' and 'feeling' can be distinguished from emotion. As Massumi (2002) puts it, emotion is subjective and, as such, it is a personal experience.

From her analysis of how a dog and its handler discover happiness in the labour of training, Haraway (2003, 2) employed the phrase "cross-species sociality" to express her notion of a mutual feeling between species. In her later work, Haraway (2008) presents animals as mindful, lively, dynamic and differentiated beings that have co-evolved with humans, and have, to draw on Latour (2004a), 'learned to be affected' by each other. However, in the article *Non-Companion Species Manifesto*, Dwyer (2007) used his perception that his son's snakes and lizards would not have mourned his son's death as evidence that Haraway's (2003) inter-species sociality was not apparent between reptiles and animals.

Further, Marseille *et al.* (2012) contend that, at zoos, the attractive charismatic, rare and endangered species, such as polar bears, are those most likely to stimulate affective reactions. Yet, through a study of recreational angling in Yorkshire rivers, Bear and Eden (2011) argue that it is not legitimate to contend that 'other' bodily forms impede inter-species affective encounters. Instead, the authors argue that affective relationships can develop between humans and fish, asserting that "fish are both affect*ed* and affect*ive*" (Bear and Eden, 2011, 341, emphasis in original). Clearly, there is an impasse as to whether non-mammalian life forms have the capacity to both affect and be affect*ed*, during leisure times and in leisure spaces. Through the study of Camilla, this chapter offers fresh insights into this debate. Before doing so, attention turns to the transition from anthropomorphic readings of animals, to more personal, egomorphic ones.

From anthropomorphism to egomorphism

Anthropomorphism is the ascription of certain *human* qualities to animals (Blok, 2007). Following Milton (2005), the use of the term 'anthropomorphic' to describe human *understandings* of animals is misleading for three reasons. First, anthropomorphism is based on assumptions made by a human about animals, for which there is little evidence in the actions and verbal statements of the animals whose understanding is being described. Second, anthropomorphism assumes that 'humanness' is the primary point of reference for understanding animals. Yet, according to Milton (2005), the *self* is the primary point of reference for such an understanding. Third, anthropomorphism implies that people understand 'things' by attributing characteristics *to* them. Alternatively, Reis *et al.* (2007) contend that humans understand things by perceiving characteristics *in* them.

Recognising the shortcomings with anthropomorphism, Milton (2005) proposes an alternative model, namely 'egomorphism'. From an egomorphic perspective, the understanding of animals as persons, that is – beings with emotions, purposes and personalities – is based on a perception of a particular animal as 'like me', in opposition to 'human-like' (Milton, 2005). From an egomorphic vantage point, I perceive emotions, purposes and personalities in animals for instance: my pet rabbit, the squirrels that run around in my garden throughout the year, or Camilla the chameleon, on the basis that they are similar to me, rather than similar to the human populace. Thus, as Reis *et al.* (2007) note, egomorphism determines that humans understand animals through their own personal experience. In short, to echo Lulka (2008), egomorphism is anthropomorphism *embodied.*

Examples referred to as 'egomorphism' in the literature are generally lacking. A notable exception is Hurn's (2008, 30) use of egomorphism in her study of the interplay of sex and gender in the commercial breeding of the horse breed, Welsh cobs, in which the author contends: "I empathize with the fact that my mare *experiences* what I recognise as broodiness". Further, there are accounts in the literature which are not explicitly referred to as 'egomorphism', yet which conform to Milton's (2005) definition of the approach. For instance, upon witnessing pets being scolded for doing 'naughty' things, such as disposing of bodily waste on a carpet, Vann (2010, 5) notes:

> I imagined what the pet was feeling by drawing from my experiences of being scolded by someone for doing troublesome things. When I got in trouble for calling my brother a mean word .... I felt embarrassed and ashamed. This experience was the closest thing I could compare to how the pet was feeling in a similar situation.

As implied above, egomorphism is a useful tool for understanding animals. Whilst somewhat neglected in the existing literature, as leisure is often rather self / ego focused, egomorphism may be a productive means of offering insight into human and animal leisure experiences. This study of Camilla therefore has potential to extend the areas to which an egomorphic understanding has been applied.

Methodology

In order to undertake the research upon which this chapter is based, I used an innovative mixed-methods qualitative approach, drawing on autoethnography and video-elicitation focus groups. According to Spry (2001), autoethnography can be viewed as the convergence of the autobiographic impulse and the ethnographic moment. Louis (1991, 365) encapsulates the spirit of autoethnography, stating: "I am an instrument of my inquiry; and the inquiry is inseparable from who I am". As such, consistent with the theory of egomorphism, when using this method I wrote personalised accounts whereby I drew on my experiences to extend understandings of Camilla's everyday life (Holt, 2003). Or as Geertz (1988, 79) puts it, I was an "I-witness" to the 'reality' I experienced. This shift towards personalised research responds to calls to place greater emphasis on the ways in which the researcher interacts with the culture being researched (Wall, 2006). By writing myself into the research, I challenge views about silent authorship (Charmaz and Mitchell 1996), in which the researcher's voice is subjugated from the presentation of research findings (Klenke, 2008). As a result, congruent with Mizzi (2010), I consider autoethnography a liberating method.

I completed my ethnographic observations over a six-week period (mid-May 2012 to end of June 2012). I typically visited Camilla five days a week, at varying times, and observed her for four hours on each occasion. In total, I undertook 120 hours of ethnographic observation. As chameleons are "just look, don't touch" reptiles (Bartlett, 2006, 2), non-participant observation was appropriate. As such, I was never directly involved in Camilla's actions, opting instead to observe them at a distance (Hall, 2008). I kept a small notebook with me and recorded my observations in this, as they occurred. To complement and enhance my ethnographic observations, I used a digital camera. However, as I was in control of the camera angles, and the moments of Camilla's daily life to capture, it must be recognised that visual imagery is always constructed: never innocent (Rose, 2007). Nonetheless, the approach was beneficial in enabling me to capture snapshots of Camilla's body language and eye contact (cf. Freeman *et al.*, 2011).

To complement my autoethnographic observations, I utilised a video-elicitation focus group method. This method was beneficial because, as chameleons are both mobile and mutable, they are difficult to capture and represent by the orthodox methodologies of the interpretative social sciences, which are concerned with the collection, interpretation and critique of disembodied representations (Lorimer, 2010). Such methods fail to appreciate the multisensory energies and intelligences of non-human bodies (Thrift and Dewsbury, 2000). Thus, to echo Whatmore (2006, 607), it is necessary to supplement the conventional palette of humanist methods, those that rely on generating talk and text, with methods aiming to amplify sensory, bodily and affective elements. Reassuringly, Lorimer (2010) argues that moving imagery has the ability to make the lively, affective animal come to the fore.

Moving imagery, despite its relative novelty as a research method, is not without criticism. For instance, like photography, videography cannot be credited with producing an objective representation of 'natural' behaviour. Everything, from choosing the camera angles during the filming stage, to deciding what material to include/exclude during the editing process, is subjective (Owens and Millerson, 2011). I used a large-screen laptop to show the video footage to focus group participants. I recruited these participants through an opportunistic sampling method, by posting requests on the Theme Park's Facebook page. I sought people who had previously visited the Creepy Critters hut, and were aware of a Chameleon residing in it. It was beneficial to have people who had seen Camilla on a previous occasion, as they could draw on prior experiences in the focus group discussion. I held two focus groups of mixed genders, each comprising eight participants. This conformed to Longhurst's (2003) recommendation of between 6 and 12 participants for a successful focus group. Each focus group took place in a quiet local coffee shop, and lasted approximately two hours. Focus groups took place here, as opposed to in the Theme Park, due to it being a quieter space, with fewer distractions. Supporting Wolch *et al.*'s (2000) opinion, focus groups allow participants to express their ideas in a relatively unconstrained manner, and to react to the opinions of other participants, thus promoting discussion and debate. Having outlined the methods upon which this research is based, I now offer an egomorphic account of Camilla's everyday life in the leisure setting.

Being Camilla: an egomorphic account

Here, I draw on Milton's (2005) notion of egomorphism, to illustrate that Camilla is not biologically and behaviourally 'other', and that Camilla, when on display to the public, is at 'work', in a human constructed leisure space. Heeding Milton's (2005) approach, Camilla's inner-world was treated as available and perceivable. As such, I drew on personal experiences, to propose ideas on what it is like to be Camilla. During one observation, I noted:

> Camilla appeared happy this morning! She came out from the leaves in the corner of her vivarium and climbed along the branches until she was on full show. Her skin was a vibrant green and her patterning was clear.

In this instance, my egomorphic reading of Camilla being 'happy' was informed by her confident mobility around the enclosure, and the dark green colour of her skin, which herpetologists take to imply a chameleon's good mood (cf. Prusten, 2008). It is noteworthy that, other than slight alterations in hue, Camilla's colour did not significantly change during the observational period of six weeks, suggesting a consistency in temperature, lighting and Camilla's mood. Nonetheless, Chameleons are known to turn red when angry or embarrassed, black when ill, and white when tired (Bartlett and Bartlett, 2001). Humans share with chameleons the ability to change colour dependent upon emotion (Changizi and Shimouois, 2011). For instance, I, supporting Darwin's (2002) observations, redden with anger, whiten with fear, display a green hue with illness, and become purplish with muscular effort. In sum, Camilla and, by extension, other chameleons, should not be pitted as *biologically* 'other' to humans.

I now argue that it is erroneous to posit Camilla as behaviourally 'other', thereby rendering problematic Martin's (1992) assertion that chameleons are 'supernatural' oddities. For instance, on one occasion, my understanding of Camilla's behaviour was informed by my familiarity with female friends 'showing off' when around groups of people. Through the following passage then, a process of "saming" (Schor 1988, 45), the counter-notion of 'othering', can be seen to have occurred:

> When entering the creepy critters' hut at 9am, Camilla was hiding behind the greenery in her vivarium. However, come 10am – when Vision Kingdom Theme Park opens to its visitors – Camilla became much more mobile. This made me think that perhaps she was aware of the presence of the crowd and, much like my human friends do adapted her behaviour in order to put on a show for her spectators. She was parading around her vivarium much to say: "don't bother looking at the snakes, all eyes should be on me!"

My egomorphic reading of Camilla 'showing off' for the benefit of Theme Park visitors, evident through the above excerpt, is comparable to Mr. MacBeth's observation of a marsh wren: "[T]he little male is the brightly little one. He calls out. He always calls for the female: 'Look at me! Look at me! Look at me! Look at me! Look at me!' 'I am so great!'" (Mr MacBeth, quoted in Reis *et al.* 2007, 151). Mr MacBeth, a teacher who uses egomorphism to facilitate his student's understanding of animal behaviour observation, and I, both draw upon egomorphism to perceive characteristics in an individual animal, in this instance: the desire for attention. Here, I suggest that Camilla shapes her activities in response to visible human presence. From this, it can be suggested that Camilla is 'working'; that is, putting on a performance for her audience in the leisure space of the Theme Park.

Further, whilst Camilla cannot articulate her thoughts or feelings using vernacular employed by humans, this is not to say that Camilla is, as Latour (2004b, 62) describes nature, "mute". The following vignette buttresses this notion:

> Camilla seemed grumpy today. On a few occasions she inflated her body and opened her mouth wide and hissed at me. This gave me the impression that she wasn't in a good mood, and that she wanted to be left alone.

As the above ethnographic vignette revealed, Camilla communicated her negative mood to me through gestures, such as the inflation of her body (see Figure 6.6 on page 98).

Likewise, when I feel grumpy, or threatened, my internal feelings are externally visible through facial expressions, and my bodily movements. To explain, when I am grumpy, I slouch and support my face with my hand, whilst when I am feeling threatened, my heart races, I breathe heavily and I become flustered. Through being attentive to Camilla's body language, I deduced that she was fed up of using her 'leisure' time performing for the public, and being subject to their constant gaze. Having provided insight into Camilla's emotions when labouring, and performing, for public consumption, I now discuss her affective capacity.

Camilla's affective capacity

The data in this section are largely drawn from video elicitation focus groups. The findings demonstrate that Camilla predominantly engenders positive emotions in visitors to the Park. Further, drawing on my ethnographic observations, I argue that human presence is capable of engendering emotions in Camilla. The majority of respondents reacted emotionally upon experiencing the video footage of Camilla, with relaxation surfacing as the most commonly evoked feeling (cf. Yerbury and Boyd, Chapter 9 this book, who discuss the relaxation gained through encounters with dolphins). Visitors to the Park regularly expressed the therapeutic affect viewing Camilla had on them. For instance, one participant noted: "having had a dispute with my wife prior to coming here over what time I'd be back for tea, I find it very soothing watching Camilla. She makes me feel at ease" (Maxwell). Despite this, Dwyer (2007, 85) is keen to remind us that "animals are inadvertent therapists". Further, for many leisure users, seeing Camilla using her rapidly extrudable tongue to devour a brown cricket engendered a feeling of fascination (see Figures 6.7–6.9 on page 99).

To quote one visitor to the Park:

> Isn't it just amazing! I really can't believe how quickly that happened! Her tongue is so long, and she just snapped up that cricket and, I mean, the cricket wasn't exactly small. Her ability to do that is just unreal, I never knew chameleons had such a cool talent [*laughs*].
>
> (William)

The above demonstrates that Camilla has potential to evoke feelings of excitement and surprise for visitors to the Park. In comparison though, some participants felt underwhelmed by Camilla. As one Park visitor put it: "she's not exactly the most exciting of animals is she. It's pretty boring watching her. All she does is walk a bit and eat" (Elaine). Here then, Camilla's inactivity and repetitive behaviour evoked a feeling of disappointment. Elaine is someone who wished to be entertained during her leisure time; but Camilla does not seem to be 'pulling her weight' on the entertainment front. It is noteworthy that, in opposition to the portrayal of chameleons in folklore, the captive chameleon did not trigger feelings of fear in the focus group respondents. These findings thus also provide a counterpoint to Markwell and Cushing's (2009, 476) generalisation that reptiles provoke "fear and anxiety".

In stark contrast to fear, for one visitor to the Park, an awareness of the pressures chameleons face in the wild stimulated feelings of love for the captive chameleon:

> In a conservation lecture at Uni we touched upon chameleons and how their habitat is being destroyed so much. Because of this, I love seeing this chameleon in captivity, enjoying itself, away from many of the pressures of the wild.
>
> (Sofia)

From this leisure user's viewpoint, Camilla is less threatened in captivity, in comparison to the wild. However, it is worth recognising that Camilla did not voluntarily give up her freedom from the wild; this freedom was curtailed for the benefit of human leisure. Equally though, whilst humans are often critiqued for invading animal spaces as part of human leisure (e.g. tourism), it is worth recognising that life in the wild can be challenging for animals (e.g. predators, sourcing food). It could thus be questioned: does Camilla have more leisure in captivity than in the wild? Does she sacrifice freedom and hard labour in the wild, for captivity and contested leisure? I would argue that Camilla's experiences in captivity are much more laborious than leisurely; she is heavily scrutinised and surveilled by Park visitors.

For another Park visitor, Camilla's presence in her enclosure evoked a negative emotional response:

> I can't help but, ya know, feel, like, sorry for the little thing. Having Karma [the participant's own chameleon], I have read quite widely about chameleons and know that the best conditions in which they flourish in is flexariums, not vivariums.
>
> (Edward)

A flexarium is a nylon mesh enclosure providing optimal ventilation for chameleons. Knowledge about the biology of chameleons resulted in this Park visitor expressing concern that Camilla's vivarium did not offer sufficient ventilation, which could hinder the reptile's everyday life. Nonetheless, whether in a flexarium or vivarium, the limited size of these facilities severely impedes the chameleon's active nature. Consequently, this visitor to the Park pitied Camilla. The diverse emotions triggered from viewing Camilla provide a counterpoint to Marseille *et al.*'s (2012) contention that attractive, charismatic, rare and endangered species are those likely to stimulate an affective reaction among zoo visitors. Findings from my research lend credence to Bear and Eden's (2011) claim that 'other' bodily forms do not preclude inter-species affective encounters. Thus far, the accounts provided depict a distinctly one-sided relationship, with Camilla being the stimulus of affect for leisure users. However, I now explore the potential for reciprocal affectivity during the leisure experience.

As with Haraway's (2003) account of dogs, the findings from this study illustrate the emotionally rewarding, yet, at times, emotionally challenging, affective interconnectivity in chameleon-human relationships. As such, the following ethnographic observation opposes Dwyer's (2007) contention that the emotional attachment humans feel for many animals is non-reciprocal:

> Camilla started the morning on the low level branches, tail curled. Her body was still, and her head moving only slightly. Camilla seemed relaxed in my presence, which made me feel relaxed too. The bond between us was mutually calming and soothing.

On the grounds that my relationship with Camilla can be described as a "joyful intersubjectivity that transcends species boundaries" (Smuts, 2001, 114), I challenge Dwyer's (2007, 74) contention that reptiles are "non-companion species". Equally, the findings, thus far, illustrate that chameleons do not warrant their literary reputations as 'supernatural oddities' (cf. Martin 1992). Instead, I posit that Camilla should be viewed as a companion animal.

However, as with close human relationships, those between people and Camilla are not only characterised by commitment and rewards, they are also shaped by ambivalence and problems (cf. Sanders, 2011). Whilst Camilla did not evoke a feeling of fear in any of the leisure users that I spoke with, it must be emphasised that Camilla was only seen on screen during the focus groups, not in person. The notion of a mutual feeling of fear between Camilla and one of her caregivers, Matt, became evident during a face-to-face encounter. Matt explained this to me during my observations:

> When we used to get Camilla out for the 'meet the creepy critters' session, parents would often take flash photos, ya know, of their kids stroking her and whatnot. But this scared Camilla. And I mean, it wasn't just Camilla that was scared, I was, and the kids were too, her facial expression made me scared she may bite me.

Although Markwell and Cushing (2009) contend that opportunities for visitors to touch reptiles can diminish the boundaries between the leisure user and the exhibited, in this instance, the opposite occurred. To expand, children typically 'stroked' Camilla – suggesting a loving, pet-like, relationship. However, on this occasion, the opportunity for tactile engagement resulted in visitors to the Park, the caregiver, and Camilla, experiencing feelings of fear. Camilla is, of course, an individual, with her own personality and temperament; it is thus worth recognising that other chameleons may have enjoyed this contact. Nonetheless, in Camilla's case, the upshot was that the boundary between visitor and exhibited became more rigid than before. Nonetheless, this section has illustrated that Camilla is both an affect*ive* and affect*ed* being; this accords with Foster's (2012) contention that the ability of reptiles to experience emotions has been underestimated. Further, through showing how fear can be shared through encounters (Keul, 2013) I have proposed that the affectivity inherent in the Camilla–human leisure experience is, in some ways, reciprocal.

Concluding thoughts

Throughout this chapter, I drew on Milton's (2005) approach of egomorphism. Egomorphism, as an approach interested in the self/ego, was useful for exploring Camilla's 'leisure' experiences. Moreover, it transpires that egomorphism is also useful for providing insight into Camilla's labour/work experiences. For example, how Camilla performs for the gaze of leisure users and her emotions experienced when doing so (e.g. happy; desired attention; or wished to be left alone). This research should thus be credited with extending the boundaries of egomorphism beyond its previous mammal and bird-centric forms. I would urge more scholars to use this empathetic methodological approach as a means of bringing to the fore the leisure and labour experiences of individual animals.

I consider that Camilla's location in a human constructed leisure space is somewhat useful for clarifying misconceptions about chameleons. Due to the spatial proximity, and daily encounter, a sense of interspecies sharing may flourish (Buller, 2012). To elaborate, the Theme Park provides a space in which people can relate to and recognise embodied similarities with Camilla; in this way, people may learn to negotiate their fears (Keul, 2013). Camilla's positioning in the Theme Park may be seen as a mutually beneficial experience – as chameleons become repositioned in the public mind (Keul, 2013). Further still, as theme parks are often associated with fun and excitement, such a leisure space may be a beneficial way of socialising people with chameleons from a young age (cf. Badr, 2014 on zoos, as a means of familiarising children with animals).

However, it is important to recognise that the placement of chameleons in captivity for human leisure may well be to the expense of the leisure life of chameleons. For instance, as argued in this chapter, Camilla experienced a feeling of fear when exposed to flash photography during a 'meet the creepy critters' session. Further, during Park opening hours Camilla's everyday life is largely subject to the gaze of Theme Park staff and visitors. Moreover, other than during reptile 'handling times', Camilla's mobility in the Theme Park is restricted, as she is confined to her vivarium. To conclude then, I argue that human leisure should be conceptualised as facilitated by Camilla's labour of presenting herself, and being subject to human gazes, day after day. Nonetheless, by considering Camilla as an affective and active agent, with the ability to shape human feelings and emotions, my research departs from much pre-existing work that considers animals as passive within leisure experiences (Carr, 2010).

Note

1 Theme Park name is a pseudonym.

References

Alexander, C. 2002. *Creating Extraordinary Joy: A Guide to Authenticity, Connection and Self-Transformation*. Alameda, CA: Hunter House.

Anderson, B. 2006. Becoming and being hopeful: towards a theory of affect. *Environment and Planning D: Society and Space*. 24 (5), pp. 733–752.

Badr, S.S. 2014. Zoos as recreational places and its impact on the Egyptian society. *Environmental Studies in Architecture and Design*, pp. 1–11.

Bartlett, R.D. 2006. *The 25 Best Reptile and Amphibian Pets*. London: Barron's Educational Series.

Bartlett, R.D. and Bartlett, P. 2001. *Jackson's and Veiled Chameleons*. London: Barron's Educational Series.

Bear, C. 2011. Being Angelica? Exploring individual animal geographies. *Area*. 43 (3), pp. 297–304.

Bear, C. and Eden, S. 2011. Thinking like a fish? Engaging with non human difference through recreational angling. *Environment and Planning D: Society and Space*. 29 (2), pp. 336–352.

Berre, F.L. 2009. *The Chameleon Handbook: Acquiring, Housing, Anatomy, Life Cycle, Health Care, Behavior, and Activities*. London: Barron's Educational Series.

Blok, A. 2007. Actor-networking ceta-sociality, or, what is sociological about contemporary whales? *Distinktion: Scandinavian Journal of Social Theory*. 8 (2), pp. 65–89.

Buller, H. 2013. Going wild in space: the porous boundaries of wild animal geographies. In Marvin, G. and McHugh, S. (Eds) *Routledge Handbook of Human-Animal Studies*. Oxon: Routledge, pp. 233–244,

Carr, N. 2010. Editorial: Animals in the tourism and leisure experience. *Current Issues in Tourism*. 12 (5–6), pp. 409–411.

Changizi, M.A. and Shimouois, S. 2011. Social color vision. In Adams, R.B., Ambady, N. and Nakayama, K. (Eds) *The Science of Social Vision*. Oxford: Oxford University Press, pp. 278–294.

Charmaz, K. and Mitchell, R.G. 1996. The myth of silent authorship: self, substance and style in ethnographic writing. *Symbolic Interaction*. 19 (4), pp. 285–302.

Darwin, C. 2002. *The Expression of Emotions in Man and Animals* [1872]. Oxford: Oxford University Press.

Dwyer, J. 2007. A non-companion species manifesto: humans, wild animals, and "the pain of anthropomorphism". *South Atlantic Review*. 72 (3), pp. 73–89.

Foster, M.S. 2012. Dealing with live reptiles. In McDiarmid, R.W., Foster, M.S., Guyer, C.J., Gibbons, W. and Chernodd, N. (Eds) *Reptile Biodiversity: Standard Methods for Inventory and Monitoring*. Berkeley, CA: University of California Press, pp. 127–130.

Freeman, C.P., Bekoff, M. and Bexell, S.M. 2011. Giving voice to the "voiceless": incorporating nonhuman and animal perspectives as journalistic sources. *Journalism Studies*. 12 (5), pp. 590–607.

Geertz, C. 1988. *Work and Lives: The Anthropologist as Author*. Stanford, CA: Stanford University Press.

Goode, D. 2006. *Playing with my Dog Katie: An Ethnomethodological Study of Dog-Human Interaction*. West Lafayette, IN: Purdue University Press.

Hall, R. 2008. *Applied Social Research: Planning, Designing and Conducting Real-World Research*. London: Macmillan Education.

Haraway, D. 2003. *The Companion Species Manifesto: Dogs, People, and Significant Otherness*. Chicago: Prickly Paradigm Press.

Haraway, D. 2008. *When Species Meet*. Minneapolis: University of Minnesota Press.

Holt, L.N. 2003. Representation, legitimation, and autoethnography: an autoethnographic writing story. *International Institute for Qualitative Methodology*. 2 (1), pp. 18–28.

Hurn, S. 2008. What's love got to do with it? The interplay of sex and gender in the commercial breeding of Welsh cobs. *Society and Animals*. 16 (1), pp. 23–44.

Jones, O. 2011. "Who milks the cows at Maesgwyn? The animality of UK rural landscapes in affective registers. *Landscape Research*. 38 (4), pp. 421–442.

Keul, A. 2013. Embodied encounters between humans and gators. *Social & Cultural Geography*. 14 (8), pp. 930–953.

Klenke, K. 2008. *Qualitative Research in the Study of Leadership*. Bingley: Emerland Group Publishing.

Krysko, K., Enge, K.M. and King, W.F. 2004. The veiled chameleon, *Chamaeleo calyptratus*: a new exotic lizard in Florida. *Florida Scientist*. 67 (4), pp. 249–253.

Latour, B. 2004a. How to talk about the body? The normative dimensions of science studies. *Body and Society*. 10 (2–3), pp. 205–229.

Latour, B. 2004b. *Politics of Nature: How to Bring the Sciences into Democracy*. Cambridge, MA: Harvard University Press.

Leitner, M.J. and Leitner, S.F. 2012. *Leisure Enhancement*. Fourth Edition. Urbana, IL: Sagamore Publishing.

Levin, A. 2015. Zoo animals as specimens, zoo animals as friends: the life and death of Marius the giraffe. *Environmental Philosophy*. 12 (1), pp. 21–44.

Longhurst, R. 2003. Semi-structured interviews and focus groups. In Clifford, N.J and Valentine, G. (Eds) *Key Methods in Geography*. London: Sage Publications, pp. 117–132.

Lorimer, J. 2010. Moving image methodologies for more-than-human geographies. *Cultural Geographies*. 17 (2), pp. 237–258.

Louis, M.R. 1991. Reflections on an interpretive way of life. In Frost, P.J., Moore, L.F., Louis, M.R., Lundberg, C.C. and Martin, J. (Eds) *Reframing Organisational Culture*. London: Sage Publications, pp. 361–365.

Lulka, D. 2008. Embodying anthropomorphism: contextualising commonality in the material landscape. *Anthrozoos: A Multidisciplinary of the Interactions of People and Animal*. 21 (2), pp. 181–196.
Markwell, K. and Cushing, N. 2009. The serpent's stare meets the tourist's gaze: strategies of display at the Australian reptile park. *Current Issues in Tourism*. 12 (5–6), pp. 475–488.
Marseille, M.M., Elands, B.H.M., and Van, M.L. 2012. Experiencing polar bears in the zoo: feelings and cognition in relation to a visitor's conservation attitude. *Human Dimensions of Wildlife: An International Journal*. 17 (1), pp. 29–43.
Martin, J. 1992. *Chameleons: Nature's Masters of Disguise*. London: Blandford.
Massumi, B. 2002. *Parables of the Virtual: Movement, Affect, Sensation*. London: Duke University Press.
Mattison, C. and Garbutt, N. 2012. *Chameleons*. London: Natural History Museum.
McCormack, D.P. 2008. Geographies for moving bodies: thinking, dancing, spaces. *Geography Compass*. 2 (6), pp. 1822–1836.
Milton, K. 2005. Anthropomorphism or egomorphism? The perspective of non-human persons by human ones. In Knight, J. (Ed.) *Animals in Person: Cultural Perspectives on Human-Animal Intimacies*. Oxford: Berg, pp. 255–271.
Mizzi, R. 2010. Unravelling researcher subjectivity through multivocality in autoethnography. *Journal of Research Practice*. 6 (1), pp. 1–14.
Neo, H. and Ngiam, J.Z. 2014. Contesting captive cetaceans: (il)legal spaces and the nature of dolphins in Urban Singapore. *Social & Cultural Geography*. 15 (3), pp. 235–254.
Owens, J. and Millerson, G. 2011. *Video Production Handbook*. Fifth Edition. Oxford: Focal Press.
Philo, C. 2005. Spacing lives and lively spaces: partial remarks on Sarah Whatmore's hybrid geographies. *Antipode*. 37 (41), pp. 824–833.
Prusten, M. 2008. High dynamic image rendering of colour in chameleons' camouflage using optical thin films. *Proceedings of SPIE*. 7057, pp. 1–8.
Reis, G., Jayme, B. and Roth, W. 2007. Egomorphism: discursive pedagogical artifact in / for science education. In Pelton, T., Reis, G. and Moore, K. (Eds) *Connections 2007*. Faculty of Education, University of Victoria, pp. 149–154. [Online], Available: http://citeseerx.ist.psu.edu/viewdoc/download?doi=10.1.1.175.2115&rep=rep1&type=pdf#page=165 165 [accessed 4 January 2017].
Rose, G. (2007). *Visual Methodologies: An Introduction to the Interpretation of Visual Materials*. Second Edition. London: Sage Publications.
Sanders, C.R. 2011. Close relationships between humans and nonhuman animals. In Arluke, A. and Sanders, C.R. (Eds) *Between the Species: Readings in Human-Animal Relations*. Paris: Pearson, pp. 45–52.
Schmidt, W. 2001. *Chamaeleo Calyptratus, the Veiled Chameleon*. Berlin: Matthias Schmidt Publication.
Schor, N. 1988. Essentialism which is not one. *Differences*. 1 (2), pp. 38–58.
Smuts, B. 2001. Reflections. In Coetzee, J.M. (Eds) *The Lives of Animals*. Princeton, NJ: Princeton University Press, pp. 107–120.
Spry, T. 2001. Performing autoethnography: an embodied methodological praxis. *Qualitative Inquiry*. 7 (6), pp. 706–732.
Stewart, M. 2008. *How do Chameleons Change Colour*. London: Marshall Cavendish.
Thrift, N. 2004. *Parables of the Virtual: Movement, Affect, Sensation*. London: Duke University Press.

Thrift, N. and Dewsbury, J.D. 2000. Dead geographies and how to make them live. *Environment and Planning D: Society and Space*. 18 (4), pp. 411–432.

Vann, J. 2010. *Introducing Animals as Living, Breathing, Thinking, Feeling*, pp. 1–28. [Online], Available: http://writing.ucsc.edu/RothmanAwards/PDFs/rothman09-10/2010-vann.pdf [accessed 4 January 2017].

Wall, S. 2006. An autoethnography on learning about autoethnography. *International Journal of Qualitative Methods*. 5 (2), pp. 146–160.

Whatmore, S. 2002. *Hybrid Geographies: Natures, Cultures, Spaces*. London: Sage Publications.

Whatmore, S. 2006. Materialist returns: practicing cultural geography in and for a more-than-human world. *Cultural Geographies*. 13 (4), pp. 600–609.

Wolch, J.K., Brownlow, A. and Lassiter, U. 2000. Constructing the animal world of inner-city Los Angeles. In Philo, C. and Wilbert, C. (Eds) *Animal Spaces, Beastly Places: New Geographies of Human-Animal Relations*. London: Routledge, pp. 71–97.

7 Human leisure / elephant breakdown

Impacts of tourism on Asian elephants

Jessica Bell Rizzolo and Gay A. Bradshaw

Introduction

Elephant riding, trekking, and circus-type shows are popular tourist activities in Asia. However, the combination of a burgeoning tourist trade and traditions that rely on the "breaking" and domination of elephants has resulted in epidemic psychological and physical breakdown in Asian elephants *(Elephas maximus)*. Captured and captive-held elephants are subjected to psychological torture, premature weaning, chronic physical abuse, and social deprivation, all of which cause profound suffering and damage. This chapter examines the relationship between elephant tourism and elephant psychosocial functioning. We show how elephant and mahout (elephant handler) wellness in the tourist industry must be examined simultaneously, as they are inextricably linked.

We begin by discussing implications of the misperception that Asian elephants are "domesticated" rather than captive-held. Then, we examine how attachment theory and traumatology, which apply equally to all vertebrate species, are ideal for conducting rigorous scientific analysis of cross-species and interspecies interactions. We illustrate these methodologies in the case of elephants and humans, and elephant–human interactions.

Next, we describe the forms and status of elephant tourism. Although elephant tourism occurs across Asia, this chapter focuses on Thailand as it has a particularly pronounced elephant tourism industry that relies heavily on captive-held elephants; it is one of the few countries where the number of elephants in captivity exceeds those in the wild. We then draw upon field research with mahouts and elephants to examine: (a) how domination-based elephant tourism leads to elephant "breakdown" (complex post-traumatic stress disorder or c-PTSD) and (b) how sanctuaries whose practices and design are informed by principles of trauma recovery, elephant prosocial values, culture, and psychology, and positive mahout–elephant attachment can serve to promote mutual wellness. Finally, we offer a framework for evaluating the sustainability of elephant tourism in Asia and offer recommendations and resources for tourists, tour operators, and tourism scholars interested in transforming the impacts of elephant tourism on Asian elephants, mahouts, and communities from exploitation to wellness.

The myth of the domesticated elephant

Humans and elephants in Asia have engaged in complex interactions for centuries; elephants have been used as labor, despised as agricultural pests, and worshipped as gods (Locke and Buckingham 2016). As a result, the centuries-old tradition of using Asian elephants both within and outside Asia has culturally conditioned many people to view captivity of this species as normative, and to consider them "domesticated." However, despite the fact that the holding of Asian elephants in captivity for logging, entertainment, and religious practices has been widespread in Southeast Asia for at least 4,000 years (Sukumar 2011), elephants remain a "wild" species.

A domesticated species is, as a result of evolutionary processes, selective breeding, or both, psycho-physiologically distinct from its wild progenitor. For example, domesticated dogs and cats are behaviorally and physiologically distinct from wolves or wild felines (Gácsi *et al.* 2005; Montague *et al.* 2014). In contrast, numerous captive-held elephants are wild-caught; those who are captive-bred are, at most, descended from wild-caught elephants. Consequently, captive-held elephants are biologically, culturally, and psychologically identical to free-living Asian elephants. While they may learn to appreciate or be forced to conform to human ways and values, captive-held elephants' culture is rooted in that of free-living elephant communities. Regardless of where they fall within the spectrum of human cultural perceptions and practices, captive and wild Asian elephants' biological and social needs remain identical.

Subsequently, to effect a scientifically rigorous evaluation of tourism impacts on elephant wellness, it is necessary to utilize free-living elephants as a baseline (Barber 2009). By definition, the analysis must include a comparison of how a captive elephant's social milieu compares with the normative attachment and socialization experiences of elephants in the wild, since that is the psychobiological context to which elephants made-captive have evolved. When speaking of whether the needs of elephants are met or unmet within elephant–human interactions, it is critical to realize that elephants are not domesticated animals.

Understanding that elephants are captive-held rather than domesticated directly informs what are viewed as "normative" conditions for elephants. Language has the power to alter perceptions of animals and to manufacture consent for particular uses of animals (Goedeke 2004; Farnsworth, Campbell, and Adams 2011; Stibbe 2001; Yates 2010). Through framing certain practices as common and inevitable, discourse can be used as a tool of legitimation and naturalization. This in turn can produce consent for such practices. For example, discourse that frames animals in circuses as "willing performers" naturalizes the notion of animals as entertainers, conceals the latent domination inherent in such practices, and endorses consent for the continuation of animals in entertainment (Bell 2015).

Elephant tourism in Thailand

The capture and use of elephants for domestic use, particularly logging, has a long history in Thailand. After logging was banned in 1989, the commercial use of elephants shifted to the tourism industry (Lair 2004). Thailand now has multiple forms of elephant tourism, including elephant riding, elephant painting (in which the elephant is forced to draw pictures for tourists), volunteering at elephant sanctuaries, and circus-type elephant shows. Tourism is Thailand's largest industry, and elephant tourism, in particular, is a growing market. As of 2009, there were a minimum of 73 Thai elephant camps; data collected a year later in 2010 counted 106 elephant tourism venues with 1,688 elephants (Nijman 2014; Schmidt-Burbach, Ronfot, and Srisangiam 2015).

Elephant tourism in Thailand is heavily biased towards entertainment rather than education. The vast majority of these venues (90.4%) provide elephant riding and almost a third (28.5%) include circus-type elephant shows. Only 3.3% of elephant tourism sites in Thailand do not use elephants for any entertainment activities (Schmidt-Burbach, Ronfot, and Srisangiam 2015). Conservation education at captive wildlife tourism sites in Thailand is virtually non-existent; 71% of these sites offer no education and only 6% offer comprehensive education (Schmidt-Burbach, Ronfot, and Srisangiam 2015).

Much of the demand for elephant riding and shows is underpinned by myths about the use of elephants in tourism. One myth is that elephants are unemployed and bored if they are not given tasks such as providing rides or performing tricks. It is also a widespread myth that elephants, being intelligent animals, voluntarily enjoy complicated, human-like tasks such as painting and dancing. Videos of elephants painting pictures with their trunk or swaying to music often go viral on the Internet, which increases tourist attraction. While elephants do enjoy play, however, they do not naturally paint figures or play organized games such as polo or soccer.

Elephants engaged in entertainment tourism sustain multiple traumas. Trauma occurs when an individual experiences overwhelming physical and/or psychological stress which he or she cannot accommodate or escape. To supply the demand for elephant tourism in Thailand, wild-caught calves and young females are illegally traded over the Thai–Myanmar border (Nijman 2014). Typically, these individuals have witnessed the death of their mothers and other family members. Whether they are captured from the wild or born into captivity, infant elephants are usually separated from their mothers and encounter numerous, prolonged traumas such as the *phajaan*, overwork, beating, severe deprivations, miscarriages due to inadequate conditions, and domination-based "management." Elephant swaying associated with music is either stereotypy, an expression of profound psychological distress and/or physical pain, or they have been physically coerced to do such on command. These are unnatural behaviors.

Similar to humans, elephants require social and mental engagement, but, again, similar to humans, only benefit when they are able to initiate activities on their own without threat and with whom and how they wish. Any normative healthful activity, mental or otherwise, occurs when elephants are unfettered in a natural environment with an ecology and society to which they have evolved.

A trans-species approach to elephant and human wellness

Theory and data have established that the neuropsychology of traumatology and attachment theory may be applied equally to all social species (humans and nonhumans) to articulate a *trans-species psychology* (Bradshaw 2005, 2009a, 2017). As such, they are ideal for conducting rigorous scientific analyses of cross-species and interspecies interactions. In other words, this framework may be used to study animals, humans, and human–animal interactions.

Trans-species psychology articulates a single model of consciousness that acknowledges the phylogenetic comparability of brain structures as well as mental and emotional capacities and consciousness among species (Low 2012). This model has been in use by science since Charles Darwin (Bradshaw 2017). Elephants and humans share common brain structures and processes that govern cognition, emotions, and other faculties. Attachment relationships are central to brain development and shared among all social species (Bradshaw and Schore 2007). For example, psychobiological structures and processes such as the hypothalamic–pituitary–adrenal (HPA) axis are comparably affected by stress and trauma among vertebrates. Just as scientific findings from nonhuman animals can illuminate the human condition, insights from human psychology and trauma offer parallel insights into the subjective experiences of nonhuman animals. This scientific symmetry is referred to as bidirectional inference (Bradshaw and Finlay 2005; Bradshaw and Sapolsky 2006).

While, conventionally, the study of elephants and nonhumans has been approached from the perspective of behavior, science's trans-species understanding compels a neuropsychological lens where behavior is subsumed as one of many expressions of psychological states. Psychology includes a spectrum of features that comprise mental experience and phenomena such as the expression of agency and emotions. Nonhuman and human animals display psychological symptoms in a number of ways. For example, elephant stereotypies (e.g., repetitive head bobbing or swaying), often described by ethologists as "purposeless" acts are, when viewed through the neuropsychological lens, extremely *purposeful* expressions reflecting profound distress (Bradshaw and Schore 2007; Bradshaw 2009b).

It is essential to compare normative elephant development, which forms the basis of their neuropsychological health, to the conditions that are standard at elephant tourist sites. In normative elephant development, calves remain in their family group for life if they are female or until adolescence (10 to 13 years old) if they are male; at this point males join an all bull group or area for a second period of socialization. Elephants are weaned on average at four years of age. Males typically do not enter *musth* until their thirties. These seemingly discrete social stages and organization are nested within a multi-tiered complex of elephant society. Hence, a synthesis of interweaving relationships shapes the infant elephant's brain and mind.

Elephants communicate through multiple media such as scent, touch, and infrasonic vocalizations (low-frequency sounds outside the range of human hearing).

These communiqués are integral to the creation and maintenance of elephant social bonds and to the socio-affective development of the calf's brain, mind, and body. The brain and mind develop in relationships and any sustained disruption has deleterious effects on physical and mental development and sense of self.

In contrast, many captive sites forcibly separate the calf from his or her mother when they are as young as one year old. When this happens, the calf exhibits severe distress, as does the mother, and struggles to be back with his/her mother. Such relational breaches affect profound trauma. Trauma experienced in infancy and childhood that disrupts attachment processes, or primary bonds between a child and caregiver, and in elephants the constellation of caregivers, *allomothers* and siblings, is referred to as *relational trauma.* A secure attachment emerges from a safe and attuned relationship between two bonded individuals. An individual who has experienced secure attachment is able to successfully regulate stress responses and emotions.

Attachment-related trauma that accrues at an early age from such events as witnessing violence towards a loved one, premature weaning and separation from attachment figure or figures, and prolonged social isolation and deprivation, lead to compromised psychological and behavioral functioning, and insecure attachment, across species. In mammals, disruption of these bonds directly influences the development of the right brain and the connections between the right prefrontal cortex and the limbic system (Schore 2001). As a result, individuals who have experienced insecure attachment (abuse, neglect, under-care) show compromised abilities to regulate stress and emotions and symptoms of hypo-arousal (e.g. emotional dissociation and withdrawal) or hyper-arousal (anxiety and unmitigated stress). In essence, the individual's ability to assess and respond to threats and to regulate stress, emotion, and social transactions is compromised. Elephants and humans are thus comparably vulnerable to psychological trauma and to c-PTSD, a condition that not only compromises an individual emotionally and physically, but also transmits across generations socially and neuro-biologically (Bradshaw and Schore 2007).

Epidemic elephant trauma

In Thailand, training young elephants to be ridden and perform tricks is commonly accomplished via the *phajaan*, or "crush" method (Cohen 2015). There are variations of the *phajaan* that depend upon the age and gender of the elephant, what actions the elephant is being trained to do, the philosophy of elephant training endorsed by the mahout, and the location of the training. However, training elephants to perform abnormal behaviors, such as riding bicycles and painting, cannot be accomplished without force. Such behaviors are forced through terror and torture by confining a young elephant in a cage only slightly larger than his/her body and inflicting relentless and unpredictable punishment using sticks, ankuses, nails, and other sharp objects until the calf submits because of physical and psychological collapse (Cohen 2015). Practices such as separating calves from their mothers, forcing elephants to ingest amphetamines to work longer hours, continuous chaining, beating, and prohibiting social interaction between conspecifics are common (Lohanan 2002; Schmidt-Burbach, Ronfot, and Srisangiam 2015). A plethora of mental and physical problems arise from such harsh environmental conditions and continual psychological duress. Eye, foot, leg, and spinal problems proliferate. Elephant spines have sharp, bony protrusions that are very sensitive to weight, and elephants ridden by tourists suffer from deformed spines and lesions. Elephants used in urban street begging experience extremely painful foot injuries from prolonged walking on hot concrete and blindness from repetitive exposure to flash photography.

Elephants are usually forbidden from touching or interacting with other elephants (see Figure 7.1). Enforced social isolation, acute anxiety, and profound fear are contrary to elephants' most basic needs. These experiences are, by definition, traumatic, and brutalizing. Elephants' natural behavior is denied, misunderstood, and punished. For example, elephants throw dirt and water on their back to protect their skin (see Figure 7.2) and touch trunks to establish and reinforce social support and love (see Figure 7.2). In captive situations where elephants are used for riding and shows, an elephant who throws dirt or water is often physically punished, as it is assumed that it interferes with the tourist's experience. Interaction between elephants is often disallowed either because of convenience, cruelty, indifference, or because these acts are misunderstood as aggression.

All of these stressors are forms of trauma (see Table 7.1) that create persistent and cumulative neuropsychological effects in elephants and other animals (Bradshaw 2009a). In cases such as elephants used for entertainment, trauma inevitably expresses as neuropsychological symptoms of c-PTSD. Psychiatrist Judith Herman created the diagnosis of c-PTSD to address conditions experienced by prisoners of war and survivors of torture and domestic violence (Herman 1997).

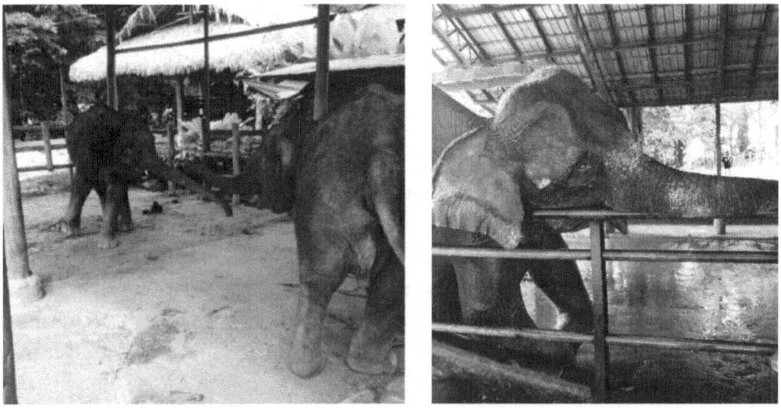

Figure 7.1 Elephants at camps often endure social frustration when they are chained and prevented from interacting with other elephants and show signs of poor welfare (left) and show signs of poor welfare (right). (© Rebecca Winkler.)

Generally, traumas associated with c-PTSD are relational (involving the disruption of social bonds), multiple and prolonged. Genesis of c-PTSD is associated with chronic, perceived and actual life-threatening environmental stress in an atmosphere of fear, unpredictability, and lack of control over self or environment, which often entails a state of physical and emotional captivity (Herman 1997). Near constant fear of harm or death, the deprivation of agency, and the inability for an individual to make his or her own life-impacting decisions are definitive characteristics of the traumas that lead to symptoms of c-PTSD. Captive-held Asian elephants are subjected to attachment ruptures such as premature weaning and maternal separation, psychological and physical torture (i.e., abuse, physical deprivation and hardship, and domination-based training designed to eliminate elephant agency), social isolation and socio-emotional deprivation, and the chronic constriction of movement and freedom (see Table 7.1); these are common factors responsible for c-PTSD development (Bradshaw 2009b).

A population-level study of c-PTSD in captive-held elephants in Thailand conducted by the authors in 2016 found a high prevalence of c-PTSD symptoms (Rizzolo and Bradshaw 2016b). The elephants we examined ranged from seven to seventy years of age and were predominantly (85%) female. Although a small percentage (4%) of these elephants were born at our observation sites (sanctuaries where natural breeding is permitted), the majority had previously been used for riding, street begging, logging, and/or circus-type shows. Our assessments encompassed structured interviews with caregivers, direct observations, and documentation of the elephants' trauma exposure (e.g. incidents of physical abuse, forced breeding, rupture of social bonds), presenting problems, medical and social history, and behavior. The elephants were assessed based on five symptom clusters (see Table 7.1) characteristic of c-PTSD: altered self-capacities, cognitive symptoms, mood disturbances, overdeveloped avoidance responses, and post-traumatic stress (Briere and Spinazzola 2005).

Overall, 74% of the elephants exhibited at least some symptoms of c-PTSD. We also looked at the prevalence of specific symptom clusters: 53% exhibited altered self-capacities, such as altered stress and emotional regulation or impaired socialization; 42% demonstrated cognitive symptoms such as overestimation of environmental danger or hyper-vigilance (for example, an elephant might exhibit

Table 7.1 Potential stressors experienced by elephants in tourism and their correlated forms of trauma

Stressor	*Form of trauma*
Premature weaning Separation from mother	Attachment rupture
Phajaan ritual	Physical and psychological torture
Social isolation	Socio-emotional deprivation
Limited nutrition	Chronic hunger
Prolonged chaining	Chronically constricted movement
Prolonged labor under physical duress/injury	Chronic physical exhaustion

severe fear or distress in response to a minor sound such as the low rumble of a truck); 53% exhibited mood disturbances such as anxiety and aggression and 38% showed avoidance behaviors such as self-injurious behavior or stereotyped behavior such as rocking, swaying, or head bobbing; 34% showed post-traumatic stress, such as distress vocalizations, avoidance, and/or a violent response to a trauma-related stimulus. The definition of "trauma-related stimulus" was based on the individual elephant's experience. Female elephants who had been forcibly bred might have this reaction to bull (male) elephants, whereas elephants who had experienced a violent mahout in the past might exhibit distress in response to the approach of any adult human male who resembled a mahout. More examples of behaviors indicative of each symptom cluster are found in Table 7.2.

An example of an elephant who exhibited numerous symptoms of c-PTSD was M.C., a seven-year-old male elephant. M.C. was separated from his mother and subjected to a version of the *phajaan* that entailed unpredictable relentless

Table 7.2 c-PTSD symptoms in a sample of captive-held Asian elephants ($N = 53$)

c-PTSD symptom cluster	*Specific symptoms*	*Examples of behavioral indicators*	*Prevalence in sample*
Altered self-capacities	Altered stress regulation Altered emotional regulation Impaired socialization	Unpredictability (abrupt altered affect) Lack of impulse control Inability to discern and communicate elephant social cues	53%
Cognitive symptoms	Overestimation of environmental danger Hyper-vigilance	Charging at mahout without provocation Flinching (expectation of cruelty) in absence of physical force Distrustful of humans Assumption that other elephants are dangerous Easily startled	42%
Mood disturbances	Anxiety Aggression	Intense social anxiety/agoraphobia (retreating/remaining in confined space away from other elephants) Physical aggression towards humans and/or elephants	53%
Avoidance behaviors	Tension reduction behaviors Avoidance of social contact	Self-injurious behavior Stereotypies Avoidance of other elephants	38%
Post-traumatic stress	Fear at trauma-related stimulus	Fear of mahouts/men Fear of bull elephants Distress vocalizations and/or violent response to trauma-related stimulus	34%

violence prior to the age of one. At the tourist camp where he was held, M.C. experienced socio-emotional deprivation, chronic hunger, and constricted movement. He was chained and isolated because he was viewed as a "problem elephant." He was starved as an attempt to punish him for his "bad behavior" and was kept chained in a yard where trash was thrown at him. The young bull was allowed less than ten minutes a day to drink water and he was beaten if he tried to put water on his back (a normative elephant behavior).

At eighteen months, he was rescued and brought to sanctuary. He continued to exhibit stereotypies, which consisted of moving his foot back and forth as though he was chained even though he no longer was. Sanctuary staff referred to this as his "invisible chain" – physically free, M.C. remained psychologically imprisoned. The stereotypy intensified during times of stress or social disruption. For example, when he had to be hospitalized for treatment of an infected tusk and when another, elder bull with whom he had bonded, died, M.C.'s stereotypy intensified. He also demonstrated unpredictable episodes of aggression towards people and other elephants, as well as fear of and aversion to water and contact with people (trauma-related stimuli). Although some of his symptoms have been mitigated through the processes of trauma recovery utilized at the sanctuary (discussed later in the chapter), they are still prevalent and persistent.

M.C.'s case is indicative of another finding from this population-level survey: relational or developmental trauma reduces resilience to other traumas and stress. In social species such as the Asian elephant, attachment, which includes both the mother–offspring bond and foundational relational experiences with other members of the family or herd, is crucial to psycho-physiological development and functioning. Deviations from this baseline environment yield trauma and lead to the suite of symptoms characteristic of relational trauma and later, if further traumatization occurs, c-PTSD (e.g. Lyons-Ruth *et al.* 2006). Practices that rupture the attachment milieu yield a traumatic stress response that reduces resilience and predisposes individuals to emotional, behavioral, and cognitive difficulties (Wilson, Hansen, and Li 2009). An individual such as M.C. who has experienced maternal separation is, from a neuropsychological perspective, poorly equipped to cope with additional traumas such as domination-based training and is therefore more likely to develop c-PTSD in the face of repetitive trauma exposure (Rizzolo and Bradshaw 2016b).

Notably, the experience of captive-held elephants finds many parallels among humans. Mahouts are also exploited in that the owners of the tourist camps receive the majority of the economic benefits of elephant riding. In many camps, mahouts receive no salary or food for the elephant; they are paid by the ride and even then receive only a small portion of the fee. It is common practice for the mahouts who live at these camps to lack adequate housing, running water, food, hygienic facilities, and medical care. Very few receive any compensation in the case of their elephant's illness or death. Most elephants in camps are forced to work long hours and receive inadequate nutrition and care until they are literally worked to death. It is not uncommon that an elephant who becomes too injured or broken down to work is left to die with no care.

This system of trans-species abuse and exploitation is maintained by ideologies of speciesism, classism, and racism towards elephants and mahouts, particularly mahouts from the Karen ethnic group, a population historically known for working with elephants (Rizzolo 2017). Mahouts have a low social status in Thailand and Karen mahouts are particularly devalued and discriminated against. A common theme expressed by mahouts is that infusing the mahout profession with economic empowerment and social status is essential for improving the mutual wellness of elephants and mahouts (Rizzolo and Bradshaw 2016a).

Elephant wellness and recovery

While c-PTSD was pervasive in the population study of captive-held elephants, elephants in sanctuary who were able to form positive social bonds with mahouts and other elephants demonstrated partial alleviation of symptoms relative to their symptom profiles upon first arrival at sanctuary (Rizzolo and Bradshaw 2016b). Critically, prosocial mahout interactions, attitude, and demeanor facilitated trauma recovery.

Although the mahout–elephant relationship can be a source of trauma for elephants when it is characterized by domination and violence, a mahout–elephant relationship based in prosocial values may serve a restorative function. Parent-child interaction is characterized by different attachment types; these are patterned ways of relating that influence the child's neuropsychological development (Benoit 2004). Given that attachment theory applies equally to all social species, such attachment types can be used as a guide for understanding mahout–elephant interactions as well (see Table 7.3).

The mahout–elephant relationship and the restoration of species-normative socialization are potentially powerful mechanisms of trauma recovery. Sanctuaries that facilitate positive social bonds between elephants by permitting elephants the agency to self-select their elephant and human companions are able to support trauma recovery in both species. This in turn promotes the reconstitution of a normative species-appropriate psychological state and hence, behavior, which provides positive feedback. For example, although the symptoms of M.C. were pervasive, they were attenuated when he began socializing with another, older bull elephant. This, in turn, supports the other elephant and also the mahout who is in charge of the elephant's care. Diminished fear in an elephant leads to diminished fear in mahouts; in neuropsychological terms, this restores the ability to regulate stress and affect. A social bond between M.C and the older bull and/or mahout can be considered a mirror of the normative all-bull socialization phase he would have experienced in the wild.

According to the model presented in Table 7.3, elephants with mahouts who were unpredictably violent (indicative of an insecure disorganized attachment style) would tend to exhibit symptoms such as contradictory behavior (e.g. aggression), stereotyped behavior, and fear. Our data suggests that this is indeed the case (Rizzolo and Bradshaw 2016a). Elephants such as M.C. who had experienced violent mahouts tended to exhibit this pattern of behavior,

Table 7.3 Application of attachment types to mahout–elephant interaction

Quality of caregiving	*Type of attachment*	*Mahout*	*Elephant*
Loving	Secure	Consistent and caring	Engaged and calm
Rejecting	Insecure avoidant	Disinterested	Withdrawn
Atypical	Insecure disorganized	Unpredictably violent	Contradictory behavior Stereotyped behavior Fear

This model also predicts that if a mahout is consistent and caring towards an elephant (characteristics of a secure attachment style), the elephant will be engaged and calm rather than fearful or aggressive. Observational data on captive-held elephants in Thailand (Rizzolo and Bradshaw 2016a) suggests that, congruent with this prediction, restorative mahout–elephant relationships demonstrated the same qualities associated with secure attachment. That is, mahout–elephant interactions modeled on secure attachment helped mitigate the symptoms of c-PTSD.

Secure attachment is characterized by a balance of security and autonomy, patience, responsiveness, and opportunities for social bonding (Benoit 2004). A caregiver promotes secure attachment by being knowledgeable about the other person's (or, in this case, the elephant's) needs and limits, and by setting limits while also providing nurturance. Restorative mahout–elephant relationships exemplified these characteristics. These mahouts provided a balance of security and autonomy by watching closely over the elephants and protecting them from danger while allowing the elephants to graze and socialize freely. Patience was on display when the mahouts, in trying to herd the elephants back to the food or medical huts, repeated verbal directions frequently and waited for the elephants to respond rather than using anger or physical force to make the elephants comply. Sanctuary mahouts created constant opportunities for social bonding through talking and patting the elephants. Their interactions were not intended to elicit any particular response from the elephants (they were not commands) but were rather for the sole purpose of building communication and trust.

Knowledge of elephants' needs and limits was demonstrated by mahouts who were highly attuned to the elephants' individual needs. For example, they would provide elderly elephants who required greater care with raincoats during the cold season or extra food. Such provisions were provided at the first sign of physical discomfort. Finally, congruent with secure attachment, these mahouts set limits when necessary but did so in a warm and flexible manner. For example, if an elephant needed to return to a particular part of the sanctuary in order to receive food or medical treatment, the mahout was firm with ensuring that the elephant followed this direction. However, the mahout was flexible with how the elephant arrived at the required destination (e.g. allowed the elephant to choose the path back) and was patient and warm while giving the elephant direction (i.e. did not become angry or violent towards the elephant).

Trauma exposure alters internal working models and socio-ecological expectations, thereby distorting an individual's sense of self and others (Siegel and Solomon 2003). By cultivating a secure attachment in sanctuary, the elephant trauma survivor is provided with a corrective emotional experience that can "retune" an elephant's neuropsychology in a manner that restores trust and emotional regulation and reduces trauma symptoms (Schore 2003). The mahout, who in other contexts may be an instrument of control and domination and hence an agent of trauma, transforms into a healer and an instrument of trauma recovery. Importantly, mahout–elephant relationships patterned on secure attachment can be healing for both elephants *and* mahouts; the relational repair and alternation of internal working models (models of self and others) is bidirectional. According to survey data from mahouts working at one elephant sanctuary (Rizzolo and Bradshaw 2016a), working at the sanctuary led mahouts to have more respect and affection for elephants. Similarly, the elephants we observed gained more trust in humans through their secure attachment to mahouts, indicating that the sanctuary environment altered the internal working models of both elephants and mahouts.

Conclusion: Is there such as thing as ethical elephant tourism?

This discussion of elephant neuropsychology and trauma is essential for conceptualizing the meaning of "sustainable" elephant tourism. A comprehensive assessment of sustainability takes into account impacts on the larger ecosystem, the economic wellness of mahouts, and the neuropsychological wellness of the elephants. Sustainability assessments of the impacts of tourism on elephants are incomplete if they are based purely on habitat and population statistics. Psychosocial criteria, such as assessments of mental and emotional status and social structure and function, are particularly important for social species such as the Asian elephant (Rizzolo and Bradshaw 2016c).

Practices that disrupt wild populations or interfere with normative socialization, such as the capture of wild calves from Myanmar for the Thai tourism trade or the widespread use of premature weaning, are mirrored in elephants' psychophysiological responses to epidemic stress and trauma. For example, work in the trans-species psychological tradition has found epidemic levels of PTSD in free-living African elephant populations subjected to years of poaching and culling (Bradshaw 2009a). Even if a group of elephants is healthy from a population point of view, in that the number of elephants is stable or increasing, there are lasting effects of psychosocial trauma and chronic stress that can affect individual and species viability. Trauma can cause a suite of disorders including behavioral aberrations (such as the inability to detect species-appropriate social signals), intra- and inter-specific aggression, infanticide, immunological compromise, increase in disease susceptibility, and depressed fertility (Bradshaw 2009a). Without inclusion of psychosocial criteria, individual and population viability and resilience are overestimated. Thus, without intervention measures to prevent trauma and to support trauma recovery, Asian elephant viability and

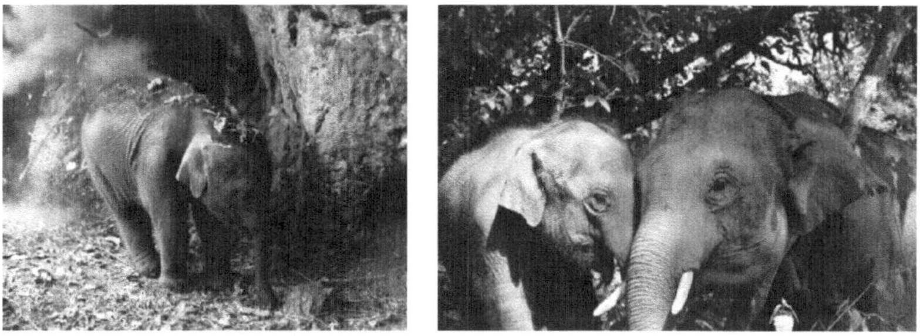

Figure 7.2 At the Mahouts Foundation in Thailand, elephants engage in natural behaviors (left) and social interaction with other elephants (right). (© Rebecca Winkler.)

resilience may drop catastrophically (Rizzolo and Bradshaw 2016c). When human fascination for elephants is harnessed to models of elephant wellness and species self-determination, as illustrated in sanctuary contexts, elephant tourism can become ethical.

Trauma recovery is possible in sanctuaries that facilitate elephants' agency (the ability to make one's own decisions), that provide opportunities for engaging in natural behaviors (Figure 7.2), and that pattern mahout–elephant interactions on the principles of secure attachment. The deprivation of agency, the inability for an individual to make his or her own life-impacting decisions, is one definitive characteristic of the traumas that lead to symptoms of c-PTSD (Bradshaw 2009b). Thus, the restoration and promotion of agency is one of the core principles of sanctuary and trauma recovery as developed by the Kerulos Center for Trans-Species Psychology. Sites that encourage elephants to develop a sense of independence, authority, and control over their lives help promote elephant self-determination, a crucial element of elephant neuropsychological health on both the individual and species level (Rizzolo and Bradshaw 2016a).

Self-determination is one of the ten principles that differentiate sanctuaries from other captive wildlife venues. These principles are knowing, safety, assurance, belonging, parity, being heard, self-determination, acceptance, support, and trust (see a description of these principles at http://kerulos.org/our-projects/being-sanctuary/10–principles-of-sanctuary/). In a sanctuary, humans are aware of the needs of the animals in their care and provide resources for their wellbeing. A sanctuary provides a place of absolute physical and psychological security and assures lifelong support for its animals (e.g. it does not remove animals for convenience or profit). Belonging and parity are created when sanctuary caregivers are receptive to the social and emotional needs of the animals and have expectations of the animal that are congruent with who the animal is and what he/she is capable of (e.g. they do not expect the animal to perform unnatural behaviors). Elephants and other animals experience being heard and self-determination when their needs are responded to in a positive, attuned manner and when they have opportunities to exercise autonomy. Self-determination is crucial for understanding why elephant venues that provide highly structured activities, such as tourists "bathing" elephants, are not sanctuaries. A sanctuary allows an elephant to choose who to be close to and what activities to engage in; this is the essence of self-determination. Forcing an elephant to be in close proximity to strangers and to adhere to a rigid schedule for human convenience (rather than for the elephant's own health or needs) is contradictory to the principles of sanctuary. Finally, a sanctuary is characterized by acceptance, support, and trust. This means that elephants are free to express their needs without fear of retribution or punishment, that they are empowered with opportunities to act on their needs and aspirations (e.g. are provided with opportunities to initiate social contact with other elephants), and that the caregiver–elephant relationship is one of trust rather than one based on domination and fear. A true sanctuary will adhere to these principles, whereas other captive venues will lack some or all of these attributes, and may rely on opposing principles. Table 7.4 provides an overview of the distinction between sanctuaries and other captive settings.

Managing tourist expectations is key to promoting elephant tourism that is congruent with the principles of sanctuary and trauma prevention and recovery. People's experience with wildlife through the virtual landscape of the media means that they approach their live encounters with animals in tourist destinations with certain preconceptions and expectations (Bentrupperbaumer 2005). The expectation of proximity to wildlife is termed the Attenborough effect. This effect creates false expectations in tourists who, after viewing animals up close for hours on television, think that their wildlife experience will include proximity to animals. These tourists may also believe that being closer to animals is the ideal wildlife experience (Fennell 2012). Tourist pamphlets that picture people riding or hugging elephants frame these activities as natural and desirable, and tend to reinforce the myth of the "domesticated" elephant who enjoys people-centered activities over natural behaviors (Rizzolo 2017). However, elephants must often be "broken" (through the *phajaan* or other methods) in order to tolerate strangers riding or being close to them. Elephant tourism sites that encourage a respectful distance between tourists and elephants (see Figure 7.3) are the most amenable to the principles of trauma prevention and recovery. Tourism resources that examine the ethical dimensions of elephant tourism and explain elephant wellness may assist in promoting tourism opportunities and expectations that are more congruent with the needs of elephants. One such tourism resource for tour operators

Table 7.4 Differentials between elephant sanctuaries and other elephant venues

Elephant sanctuary	*Non-sanctuary elephant venue*
Self-determination of elephants	Control over elephants
Companionship	Social isolation
Natural setting	Artificial environment
Exists to serve elephants	Exploitative
Trauma prevention and recovery	Yields trauma

Figure 7.3 At the Mahouts Foundation in Thailand, tourists maintain a respectful distance from the elephants. (© Peter Yuen.)

is *Tourism Cares: Creating Awareness and Respect for Elephants* (Mahouts Elephant Foundation 2015). Guides such as Horizon Guide's *Elephants in Asia, Ethically* provide a similar function for tourists (Ord 2016).

Sustainable elephant tourism must provide for the welfare needs of both elephants and mahouts. Our research suggests that, while the sanctuary model meets these criteria, standard elephant tourism yields neuropsychological trauma in elephants that profoundly impacts their emotional, social, and physiological functioning. Elephants in camps may act aggressively and have even killed tourists (Sherwell 2016). While these elephants are typically viewed as "problem elephants" and are subjected to even more punishment and social isolation after such incidents, an inability to cope with stress, hyper-vigilance, and unpredictable aggression are symptomatic of c-PTSD. Our research offers a way of thinking about elephant symptoms, such as aggression and stereotyped behavior, as not purposeless or unprovoked behaviors but rather as symptoms of trauma. From this perspective, further aggression towards these elephants will only exacerbate their aggression and will intensify the danger to tourists and mahouts. In contrast, expanding the sanctuary model would provide crucial spaces for trauma prevention and recovery that would make elephant tourism more sustainable in terms of both tourist safety and elephant wellness.

Meeting the needs of mahouts first requires understanding that mahout culture in Thailand is complex and tied to ethnic and geographic identity. There are thirteen different ethnic groups in Thailand with historical and/or contemporary elephant traditions, and these ethnic groups differ in their elephant training style, whether their elephants are used for logging, tourism or both, and their attitudes towards elephant wellness (for a complete discussion see Schliesinger 2010). For example, unlike the Kui people, the Karen people have used their elephants for logging and they train their elephants more extensively than the Kui. Karen mahouts traditionally have their elephants wander through the forest when they are not being used, whereas their Kui counterparts prefer to chain and feed their elephants close to their homes (Schliesinger 2010). According to our fieldwork with mahouts, another major distinction is between Karen mahouts, who tend to live in northern Thailand near Chiang Mai, and mahouts who train and work in Surin, Thailand. One Karen mahout noted that, unlike Surin mahouts, Karen mahouts tend to wait longer to separate elephant calves from their mothers for training and tailor the training to the individual elephant. Any interventions to discuss elephant welfare with mahouts must take in account pre-existing beliefs about elephant wellness, which often differs according to ethnicity or geography.

The mahouts we spoke to also indicated that the largest barrier to better mahout treatment of elephants is tourist demand for elephant activities, such as riding camps and circus shows that are linked to poor elephant welfare. Much of elephant mistreatment occurs because tourists desire to see elephant behaviors (dancing, painting, playing sports, etc.) that are unnatural and that can only be produced through breaking and domination-based training. One mahout who was a community leader in his Karen village noted that, "the elephants in the camps don't have good food or enough space. They stand in their own urine and feces for hours. Several of our elephants have died in the camps." In describing why he and his fellow mahouts transitioned to an elephant site that operated on a sanctuary model, he said, "We make a lot less here than in the camps. We are here because we love elephants and we want a better life for our elephants."

Overall, the mahouts indicated diverse motivations for choosing to work at an elephant sanctuary: that it is better for elephants, that it provides better work conditions and a better life for one's family, and religious considerations. Two mahouts who had used control-based methods in prior elephant-based employment said that they were surprised by how easy it was to communicate with their elephant. Both had worked with elephants for more than ten years; one had decided to work at the sanctuary because it had better work conditions, whereas the other decided to do so because it was better for elephants. Both said that since working at their current job, they have more respect for elephants and view elephants as just as intelligent as humans.

The sanctuary model can provide better work conditions for the mahouts and an improved quality of life for elephants. However, in order for sanctuaries to function, tourist demand must be substantial enough to support their existence. According to the mahouts we interviewed, tourist preferences have created the current system of domination-based elephant tourism and a shift in tourist demand is key to transforming this system. Although human leisure activities can yield a great deal of elephant suffering, human attitudes and human–elephant interactions that are patterned on a pro-social secure attachment and the principles of sanctuary have the potential to both repair and prevent Asian elephant breakdown.

Acknowledgments

Thank you to the Mahouts Foundation and Boon Lott's Elephant Sanctuary. Funding for this research was provided by the Animal Welfare Trust.

References

Barber, J.C.E. 2009. Unpacking the trunk: Using basic research approaches to identify and address captive elephant welfare concerns. In *An elephant in the room: The science and well-being of elephants in captivity*, edited by D.L. Forthman, L.F. Kane, D. Hancocks, and P.F. Waldau, 111–128. North Grafton, MA: Tufts Center for Animals and Public Policy.

Bell, J. 2015. There is no wild: Conservation and circus discourse. *Society & Animals* 23(5): 462–483.

Benoit, D. 2004. Infant-parent attachment: Definition, types, antecedents, measurement and outcome. *Paediatrics & Child Health* 9(8): 541–545.

Bentrupperbaumer, J. 2005. Human dimension of wildlife interactions. In *Wildlife tourism*, edited by D. Newsome, R.K. Dowling, and S.A. Moore, 82–112. Clevedon, NY: Channel View Publications.

Bradshaw, G.A. 2005. Elephant trauma and recovery: From human violence to trans-species psychology. PhD diss., Pacifica Graduate Institute.

Bradshaw, G.A. 2009a. *Elephants on the edge: What animals teach us about humanity*. New Haven, CT: Yale University Press.

Bradshaw, G.A. 2009b. Inside looking in: Neuroethological compromise effects in elephants in captivity. In *An elephant in the room: The science and well-being of elephants in captivity*, edited by D.L. Forthman, L.F. Kane, D. Hancocks, and P.F. Waldau, 55–68. North Grafton, MA: Tufts Center for Animals and Public Policy.

Bradshaw, G.A. 2017. *Carnivore minds: Who these fearsome animals really are*. New Haven, CT: Yale University Press.

Bradshaw, G.A. and B.L. Finlay. 2005. Natural symmetry. *Nature* 435: 149.

Bradshaw, G.A., and R.M. Sapolsky. 2006. Mirror, mirror. *American Scientist* 94(6): 487–489.

Bradshaw, G.A., and A.N. Schore. 2007. How elephants are opening doors: Developmental neuroethology, attachment, and social context. *Ethology* 113: 426–436.

Briere, J., and J. Spinazzola. 2005. Phenomenology and psychological assessment of complex posttraumatic states. *Journal of Traumatic Stress* 18(5): 401–412.

Cohen, E. 2015. Young elephants in Thai tourism: A fatal attraction. In *Animals and tourism: Understanding diverse relationships*, edited by K. Markwell, 163–179. Clevedon, NY: Channel View Publications.

Farnsworth, M. J., J. Campbell, and N.J. Adams. 2011. What's in a name? Perceptions of stray and feral cat welfare and control in Aotearoa, New Zealand. *Journal of Applied Animal Welfare Science* 14: 59–74.

Fennell, D.A. 2012. *Tourism and animal ethics*. London: Routledge.

Gácsi, M., B. Győri, Á. Miklósi, Z. Virányi, E. Kubinyi, J. Topál, and V. Csányi. 2005. Species-specific differences and similarities in the behavior of hand-raised dog and wolf pups in social situations with humans. *Developmental Psychobiology* 47(2): 111–122.

Goedeke, T. L. 2004. In the eye of the beholder: Changing social perceptions of the Florida manatee. *Society & Animals* 12(2): 99–116.

Herman, J. L. 1997. *Trauma and recovery*. New York: Basic Books.

Lair, R.C. 2004. *Gone astray: The care and management of the Asian elephant in domesticity* (4th ed.). Bangkok: FAO Regional Office for Asia and the Pacific.

Locke, P. and J. Buckingham. 2016. *Conflict, negotiation, and coexistence: Rethinking human-elephant relations in South Asia*. Oxford: Oxford University Press.

Lohanan, R. 2002. The elephant situation in Thailand and a plea for co-operation. In *Giants on our hands: Proceedings of the international workshop on the domesticated Asian elephant*, edited by I. Baker and M. Kashio, 231–238. Bangkok: FAO Regional Office for Asia and Pacific.

Low, P. 2012. The Cambridge Declaration on Consciousness. *Francis Crick Memorial Conference 2012: Consciousness in Animals*. Accessed from http://fcmconference.orrg/img/CambridgeDeclarationOnConsciousness.pdf.

Lyons-Ruth, K., L. Dutra, M.R. Schuder, and I. Bianchi. 2006. From infant attachment disorganization to adult dissociation: relational adaptations or traumatic experiences? *Psychiatric Clinics of North America* 29(1): 63–86.

Mahouts Elephant Foundation. 2015. *Tourism cares: Creating awareness & respect for elephants*. Available from www.mahouts.co.uk/image/data/blog/tourismcares.pdf.

Montague, M.J., G. Li, B. Gandolfi, R. Khan, B. L. Aken, S.M.L. Searle, P. Minx, L.W. Hillier, D.C. Koboldt, B.W. Davis, C.A. Driscoll, C.S. Barr, K. Blackistone, J. Quilez, B. Lorente-Galdos, T. Marques-Bonet, C. Alkan, G.W.C. Thomas, M.W. Hahn, M. Menotti-Raymond, S.J. O'Brien, R.K. Wilsen, L.A. Lyons, W. J. Murphy, and W.C. Warren. 2014. Comparative analysis of the domestic cat genome reveals genetic signatures underlying feline biology and domestication. *Proceedings of the National Academy of Sciences* 111(48): 17230–17235.

Nijman, V. 2014. *An assessment of the live elephant trade in Thailand*. Cambridge: TRAFFIC International.

Ord, C. 2016. *Elephants in Asia, ethically: Humane experiences with Asia's sacred animal*. Sheffield: Horizon Guides.

Rizzolo, J.B. 2017. Exploring the sociology of wildlife tourism, global risks, and crime. In *Conservation criminology: The nexus of crime, risk and natural resources*, edited by M. Gore. New York: Wiley-Blackwell.

Rizzolo, J.B., and G.A. Bradshaw. 2016a. Elephant culture, psychology, and trauma in Asia's tourism industry. Paper presented at the annual meeting of the Eastern Sociological Society, Boston, March.

Rizzolo, J.B., and G.A. Bradshaw. 2016b. Prevalence and patterns of complex PTSD in Asian elephants (Elephas maximus). In *Asian elephants in culture and nature*, edited by A. Manatunga, 291–297. Kelaniya, Sri Lanka: Centre for Asian Studies, University of Kelaniya.

Rizzolo, J.B., and G.A. Bradshaw. 2016c. Sustainability inside and out: Including psychological and social criteria for IUCN Asian wildlife assessments. Paper presented at the annual meeting of Conservation Asia, Singapore, June 2016.

Schliesinger, J. 2010. *Elephants in Thailand vol. 1: Mahouts and their culture today*. Bangkok: White Lotus Press.

Schmidt-Burbach, J., D. Ronfot, and R. Srisangiam. 2015. Asian elephant (Elephas maximus), pig-tailed macaque (Macaca nemestrina) and tiger (Panthera tigris) populations at tourism venues in Thailand and aspects of their welfare. *PLOS One* 10(9): e0139092.

Schore, A. N. 2001. Effects of a secure attachment relationship on right brain development, affect regulation, and infant mental health. *Infant mental health journal* 22(1/2): 7–66.
Schore, A. N. 2003. *Affect regulation and the repair of the self (Norton Series on Interpersonal Neurobiology)* (Vol. 2). New York: WW Norton & Company.
Sherwell, P. 2016. Elephant that killed British tourist performed at Thai safari camp tainted by cruelty allegations. *The Telegraph*, February 2.
Siegel, D. J., and M. Solomon. 2003. *Healing trauma: Attachment, mind, body and brain (Norton Series on Interpersonal Neurobiology)*. New York: WW Norton & Company.
Stibbe, A. 2001. Language, power, and the social construction of animals. *Society & Animals* 9(2): 145–161.
Sukumar, R. 2011. *The story of Asia's elephants*. Mumbai: The Marg Foundation.
Wilson, K.R., D.J. Hansen, and M. Li. 2011. The traumatic stress response in child maltreatment and resultant neuropsychological effects. *Aggression and Violent Behavior* 16(2): 87–97.
Yates, R. 2010. Language, power and speciesism. *Critical Society* 3: 11–19.

8 Volunteering for bear charities

What's in it for the bears?

Sheila Scutter and Janette Young

Introduction

Actively caring about and for animals, both domestic and wild, is a common feature of many humans lives and may involve large amounts of their leisure time (that is, unpaid) and resources (Barber 2015; Lawson, Petrovan and Cunningham 2015; Samdahl 2015). As part of caring for animals, understanding and responding to non-human needs for 'leisure' (encompassed in notions of 'enrichment' for captive animals (Young 2008)) is also engendered. Hence, in the field of animal rescue and care there is commonly an intersection of both human and animal leisure as human leisure time is spent engaged in responding to animal leisure needs. Yet this intersection has generally been ignored by researchers. This chapter focuses on these intersections with regard to one bear rescue programme funded and run primarily by volunteers, 'Free the Bears' (FTB) is based in Australia and supported by people from a wide range of countries on site at one of the sanctuaries, Phnom Tamao Wildlife Rescue Centre (National Road No. 2, Tro Pang Sap Village, Takeo Province, Cambodia, see www.wildlifealliance.org/). Funds raised by FTB volunteers based in Australia are used to support bears that have been rescued from bile farms, poachers, restaurants and illegal 'zoos' in South East Asia. Most rescued bears are unable to be released into the wild and need to be kept in captivity for the rest of their lives. Other volunteers (who may be FTB members from Australia) spend their leisure on site in Phnom Tamao, feeding bears, cleaning enclosures and providing enrichment for the bears.

Caring for animals involves more than catering for the necessities of food, water and shelter. Just as humans have other needs, as described by Maslow (1943), animals also have 'higher' needs such as safety, belonging and esteem. Burkitt (n.d.) describes how Maslow's hierarchy can be related to the needs of tigers in captivity, and the same argument can be made about providing for the 'higher' needs of bears which of necessity will remain in captivity as they are unable to look after themselves in the wild.

Bears and bear rescue

Bear species across the world are under threat as a result of environment loss, hunting for sport, extermination as pests (Islam *et al.* 2013; Vongraven *et al.* 2012;

Karamanlidis *et al.* 2014) and capture for entertainment and as pets (Burgess, Stoner and Foley 2014). Although the practices of bear-baiting and dancing bears are on the decline in some areas, they are still occurring in many countries, resulting in a life of misery for the bears involved and poaching from the wild to provide a supply of new bears for these activities (D'Cruze *et al.* 2011; Livingstone and Shepherd 2016). Bears are also killed for their meat, paws, gall bladders and skins, and kept in captivity to provide a source of bear bile, a valued product in Chinese medicine (Dutton, Hepburn and Macdonald 2011; Burgess, Stoner and Foley 2014).

Bears that have been rescued from bile farms, restaurants, poachers and traffickers, or from various forms of inhumane captivity, live out their remaining years in sanctuaries for a variety of reasons (www.freethebears.org):

- they may be cubs unable to care for themselves;
- they may have a disability or illness and need ongoing veterinary care;
- they may not have the skills to look after themselves in the wild;
- there may not be a suitable place to release them into the wild.

Many vulnerable species of bears exist in Asia, including sloth bears, sun and moon bears and pandas. Numerous organisations across the world have as part of their mission the rescue, rehabilitation, release and protection of bears. No formal listing of all bear rescue organisations has been found, but an extensive list of bear sanctuaries is provided at Bearsanctuary.com.

Bear rescue organisations based in Australia or with an Australian presence include AnimalsAsia, World Animal Protection and Free the Bears Fund. Fortunately, these organisations work cooperatively with each other and with world-wide organisations such as the World Wildlife Fund and International Animal Rescue.

The aims of these bear rescue organisations vary in detail, but generally include the following:

- to protect bears from habitat loss;
- to educate people about the plight of the bears;
- to release bears into the wild where possible;
- to rescue bears from exploitation;
- to provide a high-quality life experience for those that cannot be released back into the wild.

Volunteers contribute to bear rescue

Rescuing bears from traffickers, poachers or farms is an expensive undertaking, requiring teams of people to locate and physically retrieve and transport the bears, veterinary care, including possible surgery, and, commonly, food and support for the bears for the rest of their lives. As the aim is to provide a better life for rescued bears, spacious, stimulating and safe enclosures are required. The social needs of the bears must be attended to, as well as making their lives as much like they would be in the wild. In order to further improve bear welfare, many organisations support research into bear behaviour, nutrition, reproduction and physiology. Much of the work of these organisations is achieved by the efforts of volunteers donating their time, support and expertise to raise funds. Volunteers with bear welfare organisations can provide support in either or both of two ways; by fund-raising in their home country, or by directly assisting with the day-to-day running of the sanctuaries.

Why do people in Australia volunteer for an organisation supporting bears who live in other countries? Why bears? Why choose a charity that supports animals rather than people? Why a wild animal and not farm animals or pets? Why an animal that is not even present in Australia (the Australian native Koala-(bear) not actually being a bear – it is a marsupial; a mammal with a pouch for its young)? These are questions that Sheila has commonly been asked while making invited presentations at community group events. Sometimes the questions are underpinned with anger, as if Sheila should be supporting a more 'worthwhile' or 'important' cause.

The answers to these questions, which help us understand the motivations and experiences of volunteers, are important, as although the ultimate aim of their efforts is to improve the life of bears, if volunteers did not receive some positive outcomes, they would not volunteer and thus funds and resources for welfare organisations would disappear (Sargeant and Shang 2017). Are the motivations different for volunteers of Free the Bears Fund who raise money in Australia to send to South East Asia. compared to those who travel overseas and physically work in the sanctuary for up to six weeks? The personal costs (in terms of finances and resources of time and energy) in visiting South East Asia to volunteer for extended periods are considerable, suggesting that the positive outcomes must balance this.

Furthermore, does the volunteer activity of people in Australia make a difference to the bears – the supposedly central point of such human choices regarding leisure time and energy? What difference does the efforts of volunteers, as part of Bear Welfare organisations, make to the lives of the bears? Are the bears happier for being rescued from their previous lives? These are complex questions that go beyond the limits of this chapter, which asks the question: *What is it that volunteers observe and experience that convinces them that their efforts improve the lives of the bears in sanctuaries*? Yet these are important questions as we are faced with ever more requests for support from a wide array of charities. Given limited time and funds, knowing what it is that motivates volunteers, and what they experience as a result of their efforts, is essential to maintain their efforts (Sargeant and Shang 2017).

Methods

In order to explore the complex mesh of questions – human and animal related – raised above, a semi-autoethnography (Patton 2002) was undertaken. This kind of approach facilitates deep reflective thinking, meaning that rich qualitative data, as opposed to thin surface knowledge (Guba and Lincoln 1998), can be gathered. Explorations hinge on the experiences of one of the authors (Sheila), an Australian academic who has been involved with FTB for over 10 years, and her fellow volunteers. Sheila's involvement has been in the form of committee engagement focused on fund-raising and community education in several schools and other community forums such as Rotary and Probus in Australia, to enable freeing of bears and consequent care in several Asian countries. Hence Sheila has a wealth of expert insider or emic knowledge (Patton 2002; Young 2005) that is drawn on in this chapter and in her discussions with her fellow insiders. Sheila's co-author, Janette, is an outsider to the experiences explored in this chapter, but her outsider position provided opportunities to unpack and explore factors that as an outsider were foreign to her.

To explore the experiences and motivations for volunteers to spend their leisure time devoted to freeing bears in a foreign country Sheila contacted the Head Office of FTB. She requested permission to use their Facebook page to start a discussion on the reasons why people volunteered, what they gained from the volunteer experience and what impact they felt their volunteering had on the bears supported by FTB. This included Sheila asking to talk to/interview fellow volunteers if they were interested in discussing their involvement in FTB verbally. The FTB Office granted this permission, along with permission to use resources on their webpage. The Office also emailed past volunteers at the Phnom Tamao Sanctuary in Cambodia. This email invited people to contact Sheila to discuss motivations for volunteering in Cambodia. Interviews were held via telephone where possible (when the volunteer was in Australia) or by exchange of emails. In addition, Sheila posted on the Free the Bears Fund and the South Australian Support Group Facebook pages inviting people to contact her. At a meeting of the South Australian Support Group, members were advised that Sheila was writing a paper about motivations for volunteering with FTB and indicated agreement to her taking notes from conversations. Brief interviews were also held with individual members of the South Australian group. Notes were taken during and after discussions with groups or individual volunteers, whether face to face or by telephone. Comments on Facebook were added to these notes. Some volunteers sent written submissions and these were added to the notes. In addition, Sheila reflected on and recorded her own reasons for volunteering with FTB. The research was undertaken across 2016 and 2017.

In order to explore understandings of freed bears' leisure needs, Sheila was put in contact with the volunteer coordinators at Phnom Tamao Rescue centre (Cambodia), who provided information about the enrichment programmes offered to bears by the volunteers there and at the sanctuary in Laos. The research coordinator at Phnom Tamao also provided information about past and ongoing research taking place there.

Volunteering with Free the Bears

Free the Bears Fund originated in Perth, Western Australia, in 1993 when Mary Hutton watched a television programme documenting the plight of moon bears in bile farms in South East Asia (Hoffman 2017). Mary took to a shopping mall in suburban Perth with a petition asking the government of the day to prevent further bear farms operating (Hoffman 2017). From a small start, FTB has developed into an extensive network consisting mainly of volunteers who have had an impact on thousands of bears in Asia, India and Europe. FTB was registered as a not-for-profit charity in 1995. The achievements of FTB in working towards their four key focus areas – bear cub rescue, protecting wild bears, education and awareness, and stopping the suffering of bears – in the past 22 years are documented on their webpage at www.freethebears.org. The current focus of FTB is on the rescue of bears from Laos and Vietnam. However, most volunteers in recent years have worked at the Phnom Tamao Sanctuary in Cambodia, so this exploration encompasses mainly experiences of people who have volunteered at that location and also volunteers who are based in Australia and who raise funds to support the bears at these sanctuaries. Significantly, these volunteers in Australia may never see the animals they are working to support.

Volunteers for FTB in Australia are aligned with one of the state sub-groups and participate in a variety of fund-raising and awareness-raising activities. All such activity is organised and completed by group members, with some outside assistance as needed and as available. A major source of fund-raising is through a booth at the Royal Adelaide Show and at a variety of 'Expos' at the Wayville Showgrounds in Adelaide. Such expos may include the Animal Expo and the World Environment Expo. Interestingly, the booth at the Body, Mind and Psychic Expo is usually very successful for Free the Bears. Volunteers are organised into shifts to person the FTB booth, which sells official merchandise, donated goods, virtual gifts and sponsorships as well as accepting donations. FTB volunteers also run quiz nights, raffles and other functions, and sell entertainment books. In addition, Sheila gives talks for community groups and schools to raise awareness of the plight of the bears and the activities of FTB. People who attend these events often make donations or buy merchandise that contributes funds to the organisation.

Volunteers from many countries assist at the Phnom Tamao Wildlife Rescue Centre in Cambodia, working with the rangers to clean cages, build new enclosures and provide enrichment activities for the bears. Volunteers are provided with accommodation and some meals, and often stay for periods of around six weeks. Volunteers do not pay for the experience at Phnom Tamao, but are responsible for their own travel costs and any activities outside the sanctuary. Volunteers may also be involved in collecting data for ongoing research projects, mainly by collecting observational data for behavioural research. There is no direct physical contact with the bears, as they are described as being wild animals that can be dangerous. It is also considered not helpful to the bears to have interaction with volunteers, to avoid 'humanisation' of the bears.

It is also possible to do a one-day volunteer programme, in which volunteers produce enrichment activities for the bears by preparing and distributing food balls and by hiding foods in enclosures into which the bears are released. Volunteers pay around AUS$70 for this one-day activity, consisting of a contribution to the sanctuary and covering the cost of their supervision for the day and transport to and from Phnom Tamao. Volunteers can also participate in the educational sessions run at the sanctuary, which are focused on groups, mainly from local schools.

Emic (insider) reflections and explorations

The following is a brief summary of Sheila's engagement with FTB and her reflections on why she became, and remains involved in this project.

Sheila first heard about FTB from reading a newspaper while returning from a family holiday. There was an advert for a volunteer coordinator for the South Australian FTB support group. Although Sheila was aware of the situation of bears in bile farms, she had little understanding of the extent of the problem, and had never heard of FTB. However, something in this small unassuming advert caught her attention and she thought to herself that she would like to contribute, although she was not sure that she wanted to be a coordinator! It seems that Pauline Cockrill (who has since become a close friend and co-worker with Sheila for FTB, and was happy to be identified in this chapter) had also read this advert, and at the first meeting Sheila attended, Pauline was also there. Pauline has since been volunteering as Coordinator of FTB for the last ten years, while Sheila has volunteered in a variety of capacities including organising fund-raising events, selling tickets, purchasing and donating stock for sale and speaking at events to raise awareness and funds.

In 2014, Sheila herself participated in a one-day volunteer experience, being shown around the Rescue Centre in Cambodia. The rescue centre houses many rescued and endangered species, including bears. On this day, Sheila helped fill 'Aussie Dog' balls[1] with treats, and was able to give these to the bears by tossing them over the fence into their area. She also hid treats in a small enclosure to which the bears were later admitted, and watched them find and enjoy these delicacies. Sheila reflects:

> It was a very positive day, although I was a little disappointed not to be able to see the rescue cubs or the veterinary facilities. Seeing the bears finding the food and playing with the enrichment balls was enjoyable, but I think I would say that the thing I enjoyed most was seeing that the funds I had contributed to raising were actually being effectively used to provide good care for the bears that had been rescued.

Why do other (Australian) people volunteer with FTB?

As with so many terms, a singular definition of volunteering is elusive. However, it incorporates understandings of philanthropic motivations, care and concern for others beyond family and friends (Parker 2000) and serious 'work', despite being undertaken outside of the bounds of paid or expected employment.

The responses below are compiled from volunteers who had raised funds in Australia and those who had volunteered onsite in Cambodia. Most of those who provided input had fulfilled both roles. When reading through these notes, some clear themes and sub-themes emerged. These are presented below.

Why volunteer with animals?

'Humans can look after themselves'

Volunteers were unanimous in that the decision to contribute to an animal-based organisation (as opposed to a human-focused one) was very deliberate. Reasons for this decision varied, ranging from the feeling that there were already too many organisations supporting humans, to believing that there were too many people on the planet already. Several volunteers stated that the human race has the capacity to influence their own environment, including the communities they live in, the physical environment and the decisions made about how to live their lives. Humans can choose what to eat and how and even whether to interact with other humans. Animals, on the other hand, have very little control over their lives and were seen as powerless to resist the impact of humans without assistance. These comments related to the need to rescue animals that had been captured, and that were suffering under conditions imposed by humans. However, some volunteers also stressed the need to provide habitat by way of reserves and conservation areas, to prevent loss of habitat associated with agriculture or housing: 'We all live on this planet together, and we all have a right to exist'.

Responsibility: 'Humans destroy the environment and the animals'

It was felt that humans, by our activities, influence the experience of other animals, by destroying the environment they need, and destroying the animals themselves for entertainment, food or products such as skin, bile, fur, teeth and horns. Given that we as humans have this impact on the lives of animals, participants felt that we are responsible for addressing the situation. Volunteers also indicated that this was one of the reasons they chose to work with animals that do not exist in Australia. It was felt that the more affluent society in Australia was in a better position financially than other countries to provide the resources required. The countries where FTB operates in South East Asia (Cambodia, Laos, Vietnam) have a lower standard of living with less disposable income, which arguably impacts their ability to contribute to animal welfare. These countries are also ones where much of the poaching and trafficking of bears and bear parts occurs, so the aim of FTB in local communities is one of education about the threats to bears as a species, and the potential for employment in sanctuaries and tourism as alternatives to the use of bears in medicine.

Before it is too late

This was a continuation of the previous theme, in that if action is not taken soon to protect these animals then it will become 'too late' to save the endangered species.

There was also a sense of urgency to remove animals from cruel living conditions. Although education of local communities about the value of bears can eventually reduce the rate of habitat reduction it was feared that this will be too late for the bears. Although laws about using bear parts for traditional medicine and in local culinary dishes are emerging, there is seen to be an urgency to remove bears from potential harm as soon as possible.

Cruelty and suffering

The plight of the moon bears being kept in tiny cages and having their bile extracted on a regular basis was mentioned by nearly all volunteers. It was seen as horrific that such barbaric practices are still occurring, and hence the suffering of the bears needs to be stopped. When fund-raising at markets or community events, the suffering of bears is what volunteers are asked about most frequently by members of the public, and the major reason that many people purchase goods or make donations. Most people are aware of the bear bile farms, although they do not realise the extent of the practice across South East Asia, and volunteers provide this information while fund-raising. The volunteers felt that although the primary reason for running events was to raise money, raising awareness of the issues was also important. Some volunteers cited the way in which FTB was founded, with the efforts of one woman sending a petition to Government to stop the practice of bear farming, as being an example of the importance of raising awareness.

Endangered species

Volunteers also felt it was important that they contribute to the welfare of species that are endangered, such as bears. Here the emphasis was mainly on the sun and moon bears and sloth bears, as they are the focus of FTB and are also located in the Australasian geographic region. When asked whether other endangered species were equally important to assist, all stated that other animals (such as orangutans) were of equal importance, but that as individuals they (the volunteers) had limited time and energy and felt it better to focus on one species. However, a number of volunteers also contributed to orangutan organisations. All volunteers indicated a broader (i.e., beyond just bears) concern for the livelihoods of 'wild' endangered animals.

This said, many volunteers expressed concern about the welfare of all animals, mentioning in particular pigs and chickens as farmed animals and greyhounds as racing animals (discussions took place around the time of the expose of live baiting in greyhound training in Australia (Harazim 2016)). In addition, about 50% of volunteers indicated that they were vegetarians. Pets were included in the general concern about animal welfare, but as one volunteer said 'pets are not [a] threatened species'. It was noted that in this context pets meant domesticated dogs, cats and the like, rather than the use of bears as pets.

Individuals vs species

As mentioned earlier in the chapter, FTB has four thrusts to the work it does: *bear cub rescue, protecting wild bears, education and awareness* and *stopping the suffering*. These missions address habitat protection and education of local communities about the bears, and long-term goals that look to the future of the endangered species. Rescuing bear cubs and stopping the suffering of bears kept in bile farms and some other forms of captivity addresses the needs of individual animals and aims to give them better lives. These animals cannot be released into the wild and are unlikely to breed, so these actions will not contribute to the wild population and sustainability of the (wild) species. Protecting wild bears, and education and awareness focus more on sustaining the wild population and the species. Sheila asked FTB volunteers whether they felt that rescuing bear cubs and stopping the suffering were of equal, more or less importance than education and habitat protection for the sustainability of the species. When this topic was raised with volunteers, they were equally concerned with both. Identifying that the loss of a species for whatever reason is an indictment on the human race, as is the abuse of animals for human gain. This is an important distinction, as rescuing an individual animal and caring for it for up to 30 years is an expensive venture that does not assist in ensuring the continuity of the species.

In this regard, volunteers were conceptually trying to straddle the ethical gap identified by Singer (1993) that humans have a tendency to respond to the needs of individuals whilst often ignoring mass pain. Sheila's reflection on this tension is that while not having discussed the relative importance of the two aims, suggestions about giving priority to one over the other would evoke strong argument among volunteers. It is clear, however, that the plight of individual animals, and the changes that can be seen in their living conditions, health and well-being, is more immediately obvious to volunteers and so they tend to work towards this.

Bears' leisure

In a book focused on the rights to leisure of wild animals it is important to consider questions that pertain to bears' leisure. Do the efforts of the FTB volunteers provide or contribute to leisure for the freed bears? Do bears experience the leisure concepts of fun, enjoyment and rejuvenation (Carr in press)?

These two questions can be separated. To begin with the latter question – do bears experience leisure? Or more specifically to reflect the focus of FTB – do *captive* bears experience leisure? There would seem to be a fairly simple response to this question. No. Bears trapped in small cages that they may not even be able to turn around or even stand up in cannot be seen to be experiencing 'leisure' (AnimalsAsia 2017). On the contrary, they are brutalised. Bears moan in pain when bile is extracted and may be left to die of starvation if they do not produce enough (Kikuchi 2012). Figure 8.1 provides a less distressing visual example of bear farming than can be readily found on the internet; but clearly these bears have very little space or quality of life. The lifestyle of these caged bears is in stark contrast to the mural of bears on the far wall in the photo.

While bear farming is presented as a means of reducing hunting and killing of wild bears, there is evidence that many farmed bears are captured wild animals (Burgess, Stoner and Foley 2014; Livingstone and Shepherd 2016) and that continued demand for 'wild bear' bile (and other 'products' such as paws) has led to dual economies of wild and farmed bear bile (Dutton, Hepburn and Macdonald 2011). There are animal welfare and protection changes occurring in the region, with China reframing legislation to inhibit the consumption of endangered wildlife (Li 2014) and at least one commentator suggesting that there is a new generation of young Chinese who are leading the way in wildlife protection and care (Young 2016). However, these positive indicators do not change the current scenario which can be summarised as: Bears are endangered in the wild because of the trade in bear products (any leisure they may experience as wild animals is under threat due to risks of being captured or killed); bears are treated very poorly when commercially farmed, hence bears, especially in these imprisoned scenarios, do not have leisure.

Consequently, freeing bears can be seen to be a rights response to a sentient species. Freeing the bears provides the opportunity for these animals to seek out and experience leisure, fun and enjoyment in a way that was prohibited in their inadequate captive environments. It removes the cruel practises of being trapped in crush cages for up to 20 years, and being 'catheterised' – a medical term that does not reflect the crude practice of ramming an unsterilised hollow steel tube into a bear's gall bladder (Kikuchi 2012) – so that their bile can be drained off at regular intervals.

The efforts of FTB over 21 years is summarised on their website but includes:

- the construction and running of eight bear houses (enclosures) including every-day and specialist care such as surgery, medications and medical care;
- care of numerous bear cubs (11 in 2014 alone);
- rescue of over 500 dancing bears in India and the consequent ending of the tradition of dancing bears in that country;
- support and running of numerous education programmes in aid of bear welfare and funding of research to map wild populations of bears.

All of these undertakings and achievements (plus many more documented on the website) can be seen to be a positive addressing of the rights of bears to leisure. From this stance it does not seem an overreach to state that FTB volunteers have been instrumental in enabling those bears that they have been able to rescue to experience a high level of leisure in comparison to the commonly exploitative and abusive conditions that they have been removed from. Volunteers' actions also have implications for maintaining the rights to 'leisure' of wild bears.

For FTB and its volunteers the removal of bears from abusive conditions is not enough and their quest is to make the freedom that bears now experience as enjoyable as possible. This is because the majority of rescued bears are not able to be released into the wild; rather they remain captive, albeit a very different variety of captivity to that which they have been rescued from, and hence human dependent.

The consequent interest for FTB volunteers has become how to enable these freed bears to experience 'enrichment', or high-quality life and leisure.

Enrichment (or high-quality leisure?) for bears

Enrichment encompasses practices of seeking to enhance the quality of physical and psychological well-being of captive animals by providing them with stimulation and development opportunities similar to that which they would experience naturally in the wild (Young 2008). Patterns of interaction or avoidance with other members of one's species, seeking of sustenance and even just hanging out in the sun may all be features of wild bears' behaviour (various species differ in terms of tendencies regarding each of these features). Understandings of 'enrichment' mirror some understandings and definitions of leisure as rest, relaxation, mental stimulation and play (Carr in press).

To the outsider author of this chapter (Janette), enrichment is seeking to provide high-quality leisure. Simply releasing bears to be free to physically stretch out their bodies, to not be held captive in a space barely twice one's body size (as shown in Figure 8.1), or, as shown in more disturbing photos online, to be unable to even stand up, or sit down, or at times even to move, is to enact the right of bears to a very basic form of 'leisure'. But FTB volunteers, in line with aspirations of many modern zoos and animal sanctuaries (Young 2008) wish to add value to this basic standard. There are many guidelines available on how to provide enrichment for captive bears (see for example 'Suggested Guidelines for Enrichment' www.aazk.org/wp-content/uploads/Suggested-Guidelines-for-Bear-Enrichment.pdf and 'Sun Bears require enrichment for Jaws, Claws and Noses' https://orangutan.org/sun-bears-in-need-of-volunteer-to-coordinate-diets-and-enrichment-for-claws-jaws-and-noses/). Enrichment for bears involves seeking to replicate natural activities such as foraging. In the case of sun bears and moon bears, who are opportunistic omnivores, as their food is not readily available, they spend a lot of time seeking it, which involves exploring to find their food, and then manipulating skills to extract it. Both activities can be said to be akin to conceptualisations of leisure.

As noted earlier in this chapter, when in Cambodia, FTB volunteers contribute to enrichment for the bears by preparing treat balls or hiding food so the bears have to seek out their food. In Australia, volunteers raise funds for purchase of treat balls, hammocks and climbing frames, all of which are tools for enhancing bears' natural climbing and foraging behaviours. In thinking beyond a basic framework of leisure for the bears, enrichment combines notions of non-leisure (labour) for bears with those of play and recreation (Carr in press). So functional enrichment (a form of labour perhaps) is facilitated for the FTB bears. Do they, however, experience, or wish to experience, fun and enjoyment, or what has been termed at times hedonistic or frivolous leisure (Stebbins 2001)?

From a cross-species, physiological perspective, Berridge and Kringelbach (2008) summarise a review of the literature, stating that '[p]leasure and reward

Figure 8.1 Sun Bear Bile Extraction Operation in Mong La, Shan, Myanmar (photo: Dan Bennett, 2008).

Source: Flickr: 10, CC BY 2.0, https://commons.wikimedia.org/w/index.php?curid=13834398

are generated by brain circuits that are largely shared between humans and other animals' (p. 457). This suggests that physiological processes are in place for animals to experience pleasure akin to that experienced by humans. Behavioural research shows that replication of 'natural' activities is associated with greater well-being in animals (Young 2008) and most volunteers contributing to this chapter felt that they could tell the difference between a 'happy' bear and another by its activity, posture and appearance. In this, volunteers may well be drawing on their experiences with domesticated animals. Both of the authors of this chapter would assert that domestic animals certainly experience pleasure. Sheila can say without fear of anthropomorphising that her dog gets enjoyment out of playing with a ball and even showing off his prowess at catching a Frisbee in mid-flight. Seeing a dog or cat welcome their person home of an evening may be linked to food (if that person feeds them) but it denies non-human emotional capacities to claim that there is no pleasure in such cross-species social engagements. To do so would be to be guilty of anthropodenial, the converse of anthropomorphism; to deny animals any semblance of human emotion and engagement (de Waal 1999). Humans may be relatively good readers of domestic familiar animals, but how do humans perceive wild animals, with limited human interaction, to demonstrate fun, enjoyment and rejuvenation?

Researching the bears' leisure

Interest in maximising the quality of life of freed bears has led to FTB in Cambodia undertaking research into the enrichment activities and tools provided to the bears (Sovannarun, 2016). This research gives some indications of the levels of engagement that toys provided to bears engenders in them; engagement being perhaps a parallel to enjoyment. Bears were provided with five different toys (see Figures 8.2, 8.3 and 8.4 for examples of bears interacting with toys) and volunteers then observed the bears with each toy on four occasions for 2.5 hours on each occasion, and recorded the number of minutes the bears spent with each toy. Analysis showed that the bears spent the longest time with grass baskets, and the least time with the Aussie Dog balls and bamboo sticks. Although this data does not provide details of the quality of the interaction with each toy, it is a start in determining which toys engage the bears for the longest time. Those that offer opportunities to manipulate seem to engage the bears for the longest periods of time.

Mutually beneficial leisure?

Another way to seek to understand bear leisure is to consider reports from volunteers that perhaps reflect a mutually beneficial and pleasurable intersection of human and bear engagements. Can human volunteers themselves be part of

Figure 8.2 Bear at Phnom Tamao playing with a cube made out of webbing that contains a treat (photo: Sheila Scutter).

Figure 8.3 Bear playing with an Aussie dog ball (photo: Sheila Scutter).

Figure 8.4 Bear with a bamboo stick (photo: Sheila Scutter).

enriching the bears' lives and leisure? One of the respondents provides perhaps one of the clearest analyses of the intersections of human and animal leisure.

> Why volunteer? … It just becomes addictive! You form strong relationships with the keepers and the bears and sometimes other volunteers become good friends as well. It's great to see changes and developments in both the enclosures and the bears themselves when you're lucky like I am to go back regularly. Do the bears remember me? Possibly. Brandy was a lot more trusting and would hang around where I could talk to her a lot on my last visit and in some instances I was better at calling the bears than their keeper was! Whether that was because of familiarity I'm not sure, maybe it was just them wanting to get up their keeper's nose! As to your question about what do I think the bears get out of it You also get to know the bears on a different level from the general public, as you're much more up close and personal with them. … on my last trip, the cubs in House 2 would come running from their night dens every morning when we arrived to say hello.'

Here the intersection of this respondent's personal enjoyment in being engaged with FTB across human and bear relationships is revealed, and there seems to be some enjoyable social connections that have been made across species by some bears with the volunteer. The description is very similar to the manner in which our understandings of pet dogs' enjoyment was described earlier in the chapter; the seeking out and seemingly happy enjoyment of being in the company of a known sentient other.

Conclusion

Volunteering in aid of wild animals is a potent means of meeting the leisure needs of some humans, and, we would argue, it enhances leisure for some abused, captive wild animals. The findings of this reflective project suggest that volunteering with Free the Bears provides positive leisure outcomes for volunteers *and* for freed bears, with those bears who receive enrichment activities experiencing far more than simply freedom from extreme confinement. Bears who are freed by FTB, while generally not able to return to the wild and hence technically remaining captive, are provided with means and opportunities to experience enrichment, recreation and the 'fun' of socialising with their own and other species, including humans. All these factors can be identified as 'leisure' with regard to humans (Carr in press, Stebbins 2001) and we suggest, can be recognised as leisure for, and perhaps by, these animals as well.

Volunteers of FTB are concerned that freed bears should experience quality leisure and lives while held captive in FTB supported facilities. Looking at the photos in Figures 8.2 to 8.4 it is hard to suggest that the quality of bear leisure is not highly improved compared to that experienced by the bears in Figure 8.1. To an outsider such as the second author (Janette) the volunteers' sense of need to further enhance these animals experience is captivating. Freedom is not enough on its own. What it suggests is that FTB volunteers have a commitment to leisure as a right for these freed animals that are now in their custody.

Looking into the future, more work is required to determine how to sustain volunteer input over time – how to ensure that this leisure does not become onerous non-leisure, an unfulfilling activity undertaken in free time. This, of course, links into the existing research on volunteering (Parker 2000; Stebbins 2000, 2001; Sargeant and Shang 2017). There is also a need to explore further how to measure or identify leisure/fun/enjoyment in bears. For FTB and other agencies reliant upon volunteer activity and funding there is an ongoing need to identify how to make best use of limited resources in providing enrichment and quality leisure for bears freed of abuse, but not free to be 'wild'.

Do bears have a right to leisure? The volunteers of FTB are in no doubt. The answer is 'yes'.

Note

1 'Aussie Dog' is a company that has been supporting FTB for many years by providing balls for the bears in the sanctuaries.

References

AnimalsAsia. 2017. What is bear bile farming? Accessed 9 June 2017. www.animalsasia.org/au/our-work/end-bear-bile-farming/what-is-bear-bile-farming/overview.html

Barber, S. 2015. Nonhuman animal welfare in China: evolving rhetorical strategies for changing law and policy. *Journal of International Wildlife Law & Policy* 18 (4):309–321. doi: 10.1080/13880292.2015.1096160.

Berridge, K. and M. Kringelbach. 2008. Affective neuroscience of pleasure: reward in humans and animals. *Psychopharmacology* 199(3): 457–480. doi:10.1007/s00213-008-1099-6

Burgess, E.A., Stoner, S.S. and Foley, K.E. 2014. Brought to bear: an analysis of seizures across Asia (2000–2011). *TRAFFIC*, Petaling Jaya, Selangor, Malaysia.

Burkitt, J. n.d. The proper keeping of tigers and Maslow's hierarchy of needs. Accessed 31 May 2017. www.minoosmainecoons.com/hierarchy(1).pdf

Carr, N., in press. Meaning of leisure. In *Leisure from International Voices*, edited by K. Henderson and A. Sivan. Urbana, IL: Sagamore Publishing.

D'Cruze, N, U. Sarma, A. Mookerjee, B. Singh, J. Louis, R. Mahapatra, V. Jaiswal, T. Roy, I. Kumari, and V. Menon. 2011. Dancing bears in India: A sloth bear status report. *Ursus* 22(2): 99–105. doi: 10.2192/URSUS-D-10-00033.1.

de Waal, F. 1999. Anthropomorphism and anthropodenial: consistency in our thinking about humans and other animals. *Philosophical Topics* 27(1): 255–280.

Dutton, A., C. Hepburn, and D. Macdonald. 2011. A stated preference investigation into the Chinese demand for farmed vs. wild bear bile. *PLOS ONE* 6(7):e21243. doi: 10.1371/journal.pone.0021243.

Guba, E. and Y. Lincoln. 1998. Competing paradigms in qualitative research. In *The Landscape of Qualitative Research: Theories and Issues*, edited by Norman Denzin and Yvonna S. Lincoln, 195–220. Thousand Oaks, CA: Sage Publications.

Harazim, K. 2016. Live-baiting scandal: Disgraced greyhound racing trainers deny wrongdoing, one year on *ABC News*, 12 March. Accessed 22 May 2017. www.abc.net.au/news/2016-03-12/disgraced-greyhound-racing-trainers-deny-wrongdoing,-one-year-on/7225966

Hoffman, B. 2017. How a determined grandmother saved 900 bears. *Sunshine Coast Daily*, 9 April. Accessed 9 June 2017. www.sunshinecoastdaily.com.au/news/how-a-determined-grandmother-saved-900-bears/3164681/

Islam, M., M. Uddin, M. Aziz, S. Muzaffar, S. Chakma, S. Chowdhury, G. Chowdhury, M. Rashid, S. Mohsanin, I. Jahan, S. Saif, M. Hossain, D. Chakma, M. Kamruzzaman, and R. Akter. 2013. Status of bears in Bangladesh: going, going, gone? *Ursus* 24(1): 83–90. doi: 10.2192/URSUS-D-12-00010.1.

Karamanlidis, A., S. Pllaha, L. Krambokoukis, K. Shore, and A. Zedrosser. 2014. Preliminary brown bear survey in southeastern Albania. *Ursus* 25(1): 1–7. doi: 10.2192/URSUS-D-13-00009.1.

Kikuchi, R. 2012. Captive bears in human–animal welfare conflict: a case study of bile extraction on Asia's bear farms. *Journal of Agricultural and Environmental Ethics* 25(1): 55–77. doi: 10.1007/s10806-010-9290-2.

Lawson, B., S. Petrovan, and A. Cunningham. 2015. Citizen science and wildlife disease surveillance. *EcoHealth* 12(4): 693–702. doi: 10.1007/s10393-015-1054-z.

Li, P. 2014 China clamps down on bear farms and shark fins as animal welfare attitudes shift. *The Conversation*, 20 May. Accessed 25 May 2017. https://theconversation.com/china-clamps-down-on-bear-farms-and-shark-fins-as-animal-welfare-attitudes-shift-26936

Livingstone, E. and C. Shepherd. 2016. Bear farms in Lao PDR expand illegally and fail to conserve wild bears. *Oryx* 50(1): 176–184. doi:10.1017/S0030605314000477.

Maslow, H. 1943. A Theory of human motivation. *Psychological Review* 50(4): 370–396.

Parker, S. 2000. Volunteering as serious leisure. *Leisure/Loisir* 25(3–4): 147–155. doi: 10.1080/14927713.2000.9649912

Patton, M. 2002. *Qualitative Research and Evaluation Methods*. Thousand Oaks, CA: Sage Publications.

Samdahl, D. 2015. 'It's [not] all about the dogs': volunteers and pet rescue. In *Domestic Animals and Leisure*, edited by N. Carr, 35–66. Basingstoke: Palgrave Macmillan.

Sargeant, A. and J. Shang. 2017. *Fundraising Principles and Practice*. Second Edition. Hoboken, NJ: John Wiley & Sons.

Singer, P. 1993. *Practical Ethics*. New York: Cambridge University Press.

Sovannarun, S. 2016. Investigation into time budgets utilizing various enrichment items with captive bears. Behavioural Ecology Assignment. Royal University of Phnom Penh (unpublished).

Stebbins, R.A. 2000. Antinomies in volunteering – choice/obligation, leisure/work. *Loisir et Société / Society and Leisure* 23(2): 313–324. doi: 10.1080/07053436.2000.10707533.

Stebbins, R. 2001. The costs and benefits of hedonism: Some consequences of taking casual leisure seriously. *Leisure Studies* 20(4): 305–309.

Suggested Guidelines for Enrichment. n.d. Accessed 2 February 2017.www.aazk.org/wp-content/uploads/Suggested-Guidelines-for-Bear-Enrichment.pdf.

Vongraven, D., J. Aars, S. Amstrup, S. Atkinson, S. Belikov, E. Born, T. DeBruyn, A. Derocher, G. Durner, M. Gill, N. Lunn, M. Obbard, J. Omelak, N. Ovsyanikov, E. Peacock, E. Richardson, V. Sahanatien, I. Stirling, and Ø. Wiig. 2012. A circumpolar monitoring framework for polar bears. *Ursus* 23(sp2): 1–66. doi: 10.2192/URSUS-D-11-00026.1.

Young, J. 2005. On insiders (emic) and outsiders (etic): Views of self, and othering. *Systemic Practice and Action Research* 18(2): 151–162.

Young, R. 2008. *Environmental Enrichment for Captive Animals*. Hoboken, NJ: Wiley.

Young, R. 2016. China's wildlife in safer hands – thanks to a new generation of animal scientists. *The Conversation*, 11 February. Accessed 24 May 2017. https://theconversation.com/chinas-wildlife-in-safer-hands-thanks-to-a-new-generation-of-animal-scientists-52564

9 Wild dolphins, nature and leisure

Whose wellbeing?

Rachel Yerbury and William Boyd

Introduction

Animals regularly feature as part of people's leisure time, from relaxing at home with pets, to tourism where both captive and wild animals are encountered for human recreation. The current chapter focuses specifically on the example of wild dolphins as a part of nature, and the involvement of dolphins in human leisure activities, particularly ecotours (Curtin, 2005). This practice is supported by the human right to leisure (UNGA, 1948) as well as by economic gains (O'Connor, Campbell, Cortez & Knowles, 2009), educational and environmental benefits (Mayes, Dyer & Richins, 2004) and wellbeing advantages (Antonioli, 2005; Antonioli & Reveley, 2005; Webb & Drummond, 2001) for the human half of the interface. For this reason, timely discussion on the ethics and impacts on both sides of the human–dolphin relationship is warranted, and is the purpose of this chapter.

The chapter commences with a discussion of human leisure and the role that nature and animals occupy in human leisure, incorporating the framework of the Biophilia Hypothesis. Secondly, research considering both the benefits and negative impacts of human–wildlife interactions is examined; the beneficial argument for the wildlife ecotourism industry maintains that wildlife tourism gives people an opportunity to relax and escape, and to connect with and understand nature and animals in the natural setting (e.g. Hill, Curtin & Gough, 2013). The other point of view maintains that wild animals are used as objects and commodities, and that the focus is on human satisfaction and hedonism (Burns, 2015). The chapter then investigates further these differing positions of the human–animal dynamic; Some authors have maintained that a human-focused utilitarian approach to animals in leisure activities is restrictive and not respectful of the animal's inherent value in their own right (Wiener, 2015). Next, the chapter examines and argues for the proposition of a more ecocentric and intrinsic approach to the value of wild animals, which can provide a balanced view of interactions. Finally, how this can in turn help to guide a moral, ethical and responsible perspective and framework for leisure interactions with wild animals (e.g. Burns, 2015; Burns, MacBeth & Moore, 2011) is explored. The overall aim of this chapter is to examine the different perspectives in order to understand and evaluate the overall dynamics of the human–dolphin leisure interaction. It is argued that the non-human animal perspective needs to be considered as more than merely a commodity for the consumer (Burns, 2015). This involves not only evaluating the benefits for human wellbeing, but also exploring and understanding where dolphins fit into the interaction and ensuring their rights are considered and respected.

Leisure and wellbeing

Leisure forms an important part of people's lives, in providing a balance with other life activities and meeting human needs (Maslow, 1954), which is a necessary ingredient for psychological wellbeing. Human wellbeing refers not only to how well a person copes with life, but also how much they flourish or grow and contribute as an individual within society (Seligman, 2011). Therefore, in understanding human psychological health within a wellness, rather than a disease-based model, consideration of how people choose to spend their leisure time, is crucial (Caldwell, 2005; Haworth, 1997).

Leisure has been declared as a human right in Article 24 of the United Nations *Declaration of Human Rights* (UNGA, 1948) in recognition of the benefits it provides. Leisure time protects against mental illness by helping people to cope with stress (Iwasaki & Mannell, 2010), and difficult life events and situations (Kleiber, Hutchinson & Williams, 2010). Leisure meets various developmental (Caldwell & Witt, 2011), psychological and social needs (Tinsley & Eldredge, 1995). Positive leisure builds competencies and strengths (Passmore, 2003), and promotes flourishing (Anderson & Heyne, 2012) and positive emotional states (Hills & Argyle, 1998; Tkach & Lyubomirsky, 2006). Leisure includes opportunities to exercise freedom (Unger & Kernan, 1983) and express oneself by making choices to pursue activities that are important, meaningful and/or interesting (Ragheb, 1996). Lyubomirsky, Sheldon & Schkade (2005) maintain that up to 40% of one's long-term happiness can be accounted for by positive intentional activities, and leisure activities are examples of such purposeful behavioural choices that can have significant wellbeing benefits for humans (Caldwell, 2005; Saunders, Laing & Weiler, 2013). As much of life is restricted by obligations and schedules, leisure provides a contrast from the everyday (Mehmetoglu, 2007). This includes a chance to revitalise and reconnect with what is important to the individual (Ragheb, 1996) and the opportunity for novel or extraordinary experiences (DeMares, 2000) which can help create connections with other people, with nature and with other species, as will be discussed in this chapter.

The role of animals, specifically dolphins, in human leisure

Humans have a long history of contact with dolphins (Constantine, 2001; Orams, 1997b), a connection which is meaningful, special and even sacred for some people (Cochrane & Callen, 1992; Servais, 2005). Dolphins have formed part of the narrative and artwork of many cultures (Savage, 1990), including indigenous and ancient cultures (Neil, 2002), some of whom have long associations with dolphins in co-operative fishing (Orams, 1997b) and wisdom sharing (Jarvis, 2000).

Today, dolphins and humans have contact in many ways, including through fishing and competition for resources, in scientific research and in leisure pursuits (Allen, 2014). Dolphins remain part of spirituality as a symbol of environmentalism and health: 'dolphins have become the source of psychic healing, ecological answers, spiritual sustenance and role model for social relations' (Bulbeck, 2005, p. 84). It is suggested that the attribution of human characteristics to dolphins makes them more relatable and attractive (Jarvis, 2000; Wiener, 2015), even though these attributions may include unhelpful misrepresentations (Wiener, 2015).

People seek out contact with wildlife, including dolphins, in their leisure time (Jarvis, 2000; Wiener, 2013), with many experiences taking place via organised ecotours that involve watching, swimming with and/or feeding dolphins (Frohoff & Packard, 1995). As these experiences occur in nature, the nature effects and wildlife effects overlap and interact (Hill *et al.*, 2013). Wild cetacean tourism is a profitable industry worth an estimated 2.1 billion US dollars (O'Connor *et al.*, 2009), with over 3,300 licensed cetacean interaction operators in 120 countries around the world. As a subset of nature-based tourism, it has been discussed as a positive alternative to captive attractions (Frohoff & Packard, 1995; Hughes, 2001; Packer & Ballantyne, 2012). However, the type and impact of the wild interaction depends on the different settings and how the animals or the humans shape the interactions (Markwell, 2015). Hence, a discussion of the positive and negative impacts of dolphin ecotourism for both dolphins and humans, and the determination of the balance of impacts, is crucial (Ballantyne, Packer & Falk, 2011).

Human–nature connection: biophilia

Comprehensive research has examined the importance of human contact and connection with nature, in particular during leisure time, for healthy psychological functioning and wellbeing (for an overview, see Bowler, Buyung-Ali, Knight & Pullin, 2010). 'Nature' refers in this chapter to the natural environment, consisting of natural features, both living and non-living, rather than human-made features. It is acknowledged that the definition of nature is broad and contentious (Fine, 1997), however, such a philosophical discussion is beyond the scope of this chapter.

A frame within which to consider how leisure time in the natural environment affects wellbeing, is the Biophilia Hypothesis (Wilson, 1984). This framework posits that wellbeing is affected by an individual's connection or disconnection with nature and that humans experience psychological disequilibrium and problems when they lose touch with nature (Kellert, 1996). The historic human move from rural to urban environments, in many parts of the world, has meant a move away from natural environments and into increasingly urban and technology-based cities, which has decreased human contact with nature and animals (Pergams & Zaradic, 2008). Leisure time provides an opportunity to re-engage with nature and animals, and to gain wellbeing benefits from these interactions.

Human–dolphin interactions and human benefits: wellbeing

As part of searching for connection with nature, humans crave emotional connection to animals to defeat feelings of isolation from nature (Kellert, 1996; Myers & Saunders, 2002). Studies suggest that close recreational contact with wild animals adds to the psychological benefits of the nature experience (Besio, Johnston & Longhurst, 2008; Cloke & Perkins, 2005; Curtin, 2005, 2009; Fiedeldey, 1994; Muloin, 1998; Schänzel & McIntosh, 2000; Wiener, 2013). In particular, dolphin interactions have the ability to uplift people's emotional states (Cloke & Perkins, 2005; Curtin, 2006; DeMares, 2000; Milstein, 2008), create nature connections (Besio *et al.*, 2008; Wiener, 2013) and decrease people's levels of anxiety and stress (Webb & Drummond, 2001). People have reported intense feelings, such as happiness, as well as emotional release and relaxation when they interact with wild cetaceans including dolphins, porpoises and whales (Cochrane & Callen, 1992; Curtin, 2009; Dobbs, 2000). These intense experiences have been discussed in terms of meaningful 'peak experiences' (DeMares, 2000; Webb & Drummond, 2001) and awe (Birtles, Valentine, Curnock, Arnold & Dunstan, 2002).

In her phenomenological study, Curtin (2006) found that participants who swam with dolphins had feelings of being uplifted and satisfied, and reported it to be an important experience in their lives. This was more the case for participants swimming with wild dolphins than those swimming with captive dolphins. This study concluded that experiencing animals in their natural habitat provides a greater significance for people, due to authenticity and cognitive consonance with beliefs (Curtin, 2006). Wellbeing effects are determined in part by tourist perception and satisfaction, which is influenced by a number of factors including length and intensity of interaction, and emotional responses to the interaction (Jarvis, 2000; Muloin, 1998; Valentine, Birtles, Curnock, Arnold & Dunstan, 2004). Jarvis (2000) examined wild dolphin encounters in terms of proximity and interaction and found that being in the water allowed ecotourists to have a closer and more intimate view of the dolphins, with a chance of eye contact or experience of sonar, 'people felt moved by their experience of the perceived communion with the animal and the opportunity to become momentarily and completely immersed in the world of the other' (p. 244).

Undoubtedly, there are wellbeing benefits for humans in interacting with cetaceans in leisure time. However, in order to maintain a complete and balanced view of human–dolphin leisure interactions, the effects on dolphins also need to be explored.

Human–dolphin interactions and human benefits: social and environmental benefits

Along with the human wellbeing benefits, there are other valuable impacts of human and dolphin leisure interactions, which could have a flow-on effect to benefit dolphins as well as humans. According to Duffus and Dearden (1990) in an early discussion of ecotourism, 'the actual total value of wildlife involves a wide spectrum of market and non-market values' (p. 214). Similarly, in their review of positive and negative impacts of wildlife ecotourism, Green and Higginbottom (2000) discuss how wildlife tourism can include potential nature and wildlife benefits such as financial benefits for nature via projects funded by tourist contributions (such as park use fees) or practical projects funded or undertaken by ecotour operators. Economic incentives for nature and wildlife conservation, as well as visitor awareness and education leading to conservation behaviours are also potential positive effects (Green & Higginbottom, 2000).

In the specific case of dolphin ecotourism, many studies highlight the benefits of increased nature connection and awareness facilitating attitudinal and behavioural change (e.g. Kortenkamp & Moore, 2001; Mayes *et al.*, 2004; Orams, 1997a; Zeppel, 2008; Zeppel & Muloin, 2008). Visitor education research demonstrates that as humans experience and learn more about marine mammals, they increase their connection, value and positive attitudes towards them (Mayes *et al.*, 2004). This has the opportunity to create mutual benefit: 'a reciprocal relationship may exist between the natural environment and the people who engage with it: experience of the natural environment may be able to simultaneously promote affective wellbeing on the one hand and pro-environmental orientations on the other' (Hinds & Sparks, 2009, p. 185). In particular, wild dolphin tourism has been shown to provide a more effective vehicle for tourist attitude change than captive experiences (Curtin & Wilkes, 2007; Packer & Ballantyne, 2012) whilst incorporating fewer negative impacts and ethical issues than captive dolphin leisure interactions (Hughes, 2001).

Positive ecotourism outcomes for dolphins are influenced by both internal and external factors (Moscardo, Woods & Saltzer, 2004). In terms of internal factors, the visitor's prior experiences (Neil, Orams & Baglioni, 1996), beliefs (Amante-Helweg, 1996) and expectations (Mayes *et al.*, 2004), as well as the emotional connection formed with the animal (Ballantyne *et al.*, 2011; Milstein, 2008; Neil *et al.*, 1996; Zeppel & Muloin, 2008) influence whether an individual's attitudes and/or behaviour will be impacted by the experience. Externally, the quality of the ecotour experience (Birtles *et al.*, 2002) and the interpretation provided on the tour (Mayes & Richins, 2009), are also determining factors in lasting attitudinal and/or behavioural change (Orams, 1997a; Orams & Hill, 1998). Lück (2003) found that ecotourists often want more education about marine mammals and marine environments than is provided on tours, and Stamation (2008) found that 76% of boat-based whale watchers rated learning about whales as important. Mayes *et al.* (2004), however, report that seeking education or knowledge motivated fewer than half of visitors to two different Australian wild dolphin interaction sites. Despite this, Mayes *et al.* (2004) found that highly managed and structured interpretative programmes did have an impact on subsequent visitor intent for pro-environmental behaviour. Furthermore, a number of researchers maintain that high-quality education needs to incorporate cognitive reflection and emotional connection to wildlife, particularly empathy, in order to facilitate change (Ballantyne *et al.*, 2011; Neil *et al.*, 1996; Orams, 1994; Zeppel, 2008; Zeppel & Muloin, 2008). Therefore, produced from a human-based viewpoint, with potential to be mutually beneficial, these effects are only possible under certain circumstances, with the human as the determiner of the effects. In terms of ensuring positive impacts for both humans and dolphins, attention needs to be given to high-quality interpretation programmes which create lasting attitudinal and behavioural change (Mayes *et al.*, 2004; Mayes & Richins, 2009).

Human–dolphin interactions: benefits disrupted

There are many documented negative impacts on wildlife in leisure activities with humans (e.g. Allen, 2014; Green & Higginbottom, 2000). Dolphin ecotourism has recently been described as sitting in the middle of the conservation and commercial ends of the continuum (Wearing & Jobberns, 2015), based on the human–wildlife interaction continuum by Duffus and Dearden (1990). Some argue that cetacean ecotourism is still commercialist and commodification based (Wiener, 2015), is capitalist in character (Higham & Neves, 2015) and is harmful to the wildlife (Allen, 2014; Bejder *et al.*, 2006). Others, however, maintain that wildlife tourism should be defined as non-consumptive as the death of the animal is not the intention of the activity (Burns *et al.*, 2011), but there is also recognition that the distinction is blurry (Fennell, 2013a). While it is acknowledged that well-managed and regulated cetacean tourism is a preferable option to cetacean hunting and exploitation in captivity, ecotourism is not a benign activity (Allen, 2014; Bejder *et al.*, 2006; Higham, Bejder & Williams, 2014). In particular, poor management and regulatory non-compliance on dolphin expeditions can contribute to an array of negative issues such as illness, injury and mortality, pollution, habitat destruction or modification, as well as stress and behavioural changes (Orams, 1997a; Spradlin, Devenak, Terbush & Nitta, 1999). Even well-managed ecotours inadvertently add traffic, noise and pollution to the habitat (Allen, 2014; Richardson & Würsig, 1997; Steckenreuter, Harcourt & Möller, 2012). Blewitt (2008) discusses the significant effects of ecotour and recreational boats on dolphin behaviour. Such behavioural changes include short-term changes of direction and surface activities, and decreases in feeding and resting behaviour (Constantine, Brunton & Dennis, 2004), which potentially lead to longer-term population changes. Additionally, the evidence for tolerance of, and habituation to, humans by dolphins has been raised as a risk factor by increasing vulnerability to danger (Frohoff & Packard, 1995; Martinez, Orams & Stockin, 2011). Long-term studies (e.g. Bejder *et al.*, 2006) have also reported a decrease in abundance of dolphins exposed to tour boat operations, reinforcing the idea that wildlife ecotourism is not a harmless activity. Similarly, other researchers (Constantine, 2001; Constantine *et al.*, 2004) agree that avoidance and disturbance behaviours occur in response to long-term exposure of bottlenose dolphins to ecotours.

In particular, provisioning of wild dolphins, even as part of well-managed ecotourism, has been shown to have a significant impact on individuals and populations (Orams, 1995, 2002; Orams, Hill & Baglioni, 1996). The potential impacts for dolphins involved with feeding programmes include an increase in dependency and habituation (Orams, 2002), where animals rely on humans and become less effective at self-feeding and more likely to approach and seek interaction with humans in general (Finn, Donaldson & Calver, 2008; Smith, Samuels & Bradley, 2008). Being generally less wary of humans can lead to increased susceptibility to boat strikes and other injuries (Samuels & Bejder, 2004) as well as exposure to illness, disease and injury, higher infant mortality and reduced reproductive success (Mann, Connor, Barre & Heithaus, 2000). Reductions in behaviours such as resting, feeding and maternal care (Foroughirad & Mann, 2013) and increases in begging and aggression can also result from provisioning (Orams *et al.*, 1996; Scheer, 2010).

A less obvious, yet fundamental, negative impact of dolphin ecotourism is the tourist perception of nature and dolphins, often informed by industry and media representations (Besio *et al.*, 2008; Wiener, 2015), which can portray dolphins as objects or commodities (Burns, 2015). Such depictions form the basis for the way that wildlife is viewed, valued and ultimately treated and/or used in leisure interactions. Some representations may be helpful in conservation; for example, dolphins being represented as both peaceful, intelligent, yet vulnerable and in need of protection (Cloke & Perkins, 2005). However, industry metaphors of 'nature as performer' (Milstein, 2015) with perpetually happy dolphins seeking out human interactions (Wiener, 2015), and always ready to perform (Cloke & Perkins, 2005), have promoted tourism with a human-centred focus, without adequate regard to the wildlife impacts. Furthermore, concern has been expressed about the ecotour narratives of dolphins as both 'wild' and 'domestic' for the purpose of appealing to customers (Besio *et al.*, 2008). Such constructions of dolphins using appealing and attractive misrepresentations to create and sell a product (Wiener, 2015) strengthen the human- centred view of humans as separate to nature and more important and powerful than animals (Bekoff, 2008; Hill *et al.*, 2013).

Human–dolphin leisure interactions: anthropocentric versus ecocentric perspectives

In the environmental ethics and politics literature (e.g. Eckersley, 1989), human–nature relationship positions range from a human-centred or anthropocentric stance, to a systemic approach of ecological consciousness or ecocentrism (Leopold, 1989). This continuum of moral reasoning is a useful way to consider how people value and conceptualise and therefore interact with nature and wildlife (Kortenkamp & Moore, 2001). These positions are in part determined by the way the media and industry portray and interact with dolphins, the messages given to the public and the way these messages are interpreted (Wiener, 2013). Underpinning this conceptualisation, is the way in which research, legislation and the leisure industry refer to and protect dolphins via best practice (Garrod & Fennell, 2004; O'Neill, Barnard & Lee, 2004). For example, animals and their rights are missing from the *World Tourism Organization's Global Code of Ethics for Tourism* (UNWTO, 1999) and are often not afforded recognition in leisure encounters. In line with this, Fennell (2013b) maintains that the use of ecocentrism as a guiding theory in animal tourism has not been widely undertaken.

Some argue that wildlife tourism can be seen as largely hedonistic (Burns, 2015) and anthropocentric, due to its focus on the best experience for the human customer (Burns *et al.*, 2011), regardless of nature impacts or rights. In his writing on the ethics of dolphin–human contact, White (2007) maintains that dolphins have both inherent value and moral standing due to their sentience, self-awareness and superior consciousness. These factors, he argues, afford dolphins intrinsic worth, recognition and rights. This, he continues to argue, generates human obligations to cease engaging in interactions that have negative or even questionable outcomes for dolphins (White, 2007). According to Fennell (2013b), there are various levels on the continuum between anthropocentrism and ecocentrism, depending on the extent to which human benefit is the core focus of the position. An extreme ecocentric response to issues surrounding dolphin ecotourism, for example, may result in the advocacy of no leisure contact at all; a response designed to prevent any potential harm to the animals and the ecosystem. This aligns with the preservationist views of early conservationists such as John Muir and the untouched wilderness perspective of the 1970s and 1980s (for a discussion see Gomez-Pompa & Kaus, 1992; Turner, 2002). A less extreme response (though still ecocentrically influenced) may include allowance of monitored and regulated activities within parameters, in recognition of potential mutual benefits to humans, wildlife and the environment. In this case, an ecocentric approach to wildlife ecotourism allows a reconsideration of the relationships between humans and dolphins within the context of active inter-species engagement and acknowledges the interdependence of humans and other life (Milstein, 2015). Under such conditions, a seven-factored ecocentric model for wildlife tourism has been proposed (Burns *et al.*, 2011), in which wildlife is respected for its own inherent value in nature as part of the interconnectedness of all life. This incorporates acknowledging a moral obligation to protect, respect and do no harm within properly managed activities (Burns *et al.*, 2011).

In line with the responsible continuation of leisure activities, a number of researchers have made practical recommendations about the management of wild cetacean populations in the context of leisure interactions (e.g. Lusseau & Higham, 2004; Martinez *et al.*, 2011; Mayes & Richins, 2009; O'Neill *et al.*, 2004; Stamation, 2008; Zeppel, 2009). These suggestions acknowledge the rights of wildlife and include minimal disturbance of cetaceans (O'Neill *et al.*, 2004; Valentine *et al.*, 2004), incorporating the importance of swimmer placement (Martinez *et al.*, 2011), areas of designated marine sanctuaries (Lusseau & Higham, 2004), time-limited encounters, no approach times, and limited boats in the area (Stamation, 2008; Zeppel, 2009). Many researchers agree that these strategies need to include monitoring by regulatory services for the adherence to the marine mammal legislations and guidelines (Martinez *et al.*, 2011; Neil *et al.*, 1996; Zeppel, 2009), with increased presence and jurisdiction (Stamation, 2008), and a code of conduct which is internationally recognised (Garrod & Fennell, 2004).

Effective tour education (O'Neill *et al.*, 2004; Orams, 1996; Orams & Hill, 1998) and ongoing research (Frohoff & Packard, 1995; Stamation, 2008) are important aspects of ensuring that the rights and welfare of cetaceans are met. Of course, if the ecocentric argument is taken to its extreme extent, the possibility of no contact needs to be acknowledged as a possible course of action. The strategies listed above, however, represent a middle ground between complete exploitation of cetaceans and complete cessation of contact with cetaceans, and can result in a balanced, ecocentric approach of mutual benefit.

Changing perceptions of cetaceans so they are seen as *inherently* valuable, can also have a significant impact on whether people view cetacean leisure interaction from an anthropocentric or an ecocentric position (Burns *et al.*, 2011; Wiener, 2015). Research into the value of cetaceans, conducted by Kellert (1999) and Kellert, Gibbs and Wohlgenant (1995) found that more moralistic and humanistic values were held by younger and more educated people who were more likely to be against cetacean exploitation, fitting with an ecocentric position. The value system depends heavily on how nature and the wildlife experience are constructed by media and the industry; 'reformulating the tourist as witness—instead of audience/fan/talent scout—could help shift watchfulness of wildlife watching from an overly simplistic anthropocentric aperture to that of ecological interdependence and informed restorative responses.' (Milstein, 2015, p. 244).

Conclusion

Humans and dolphins have a long history of interaction, which creates both positive and negative impacts. The importance of such interactions for humans are encompassed in the Biophilia Hypothesis, which highlights the need to connect with nature and wildlife in order to ensure human psychological wellbeing. Humans therefore have a right to leisure and to improve their wellbeing by interacting with nature and wildlife. An exclusively human or anthropocentric position, however, disallows a comprehensive and balanced view of the interface. The negative impacts on dolphins and nature need to be understood and acknowledged to allow a deeper critique and remediation of harm. This will support a move beyond the current situation characterised by negative impacts for cetaceans currently sanctioned by human entitlement and unawareness. Although there are potential positive impacts for cetaceans in human–dolphin interactions, the ecocentric position enables the wellbeing of the dolphin to be considered not just as a commodity for human benefit, but intrinsically in its own right. Refocusing with an ecocentric viewpoint allows a deeper critique and thus potential reconsideration of the encounters and relationships between dolphins and humans. This allows for a depth of concern for the welfare of the dolphins in their own terms, not human terms. Such re-evaluation highlights many possible ecocentrically focused courses of action, from complete cessation of contact, to a continuation of human–dolphin interaction within a new paradigm of human–dolphin engagement. On a wider level, this also has bearing on the application of research to legislation, best practice models and compliance regulation and monitoring. If leisure experiences are positioned ecocentrically, they produce an opportunity to understand and value nature and wildlife within the context of an interconnected living system. Within such a framework, both sides of the equation are considered and the likelihood of a mutually beneficial relationship between humans and cetaceans will increase.

References

Allen, S. J. (2014). From exploitation to adoration: the historical and contemporary contexts of human-cetacean interactions. In J. E. S. Higham, L. Bejder & R. Williams (Eds), *Whale Watching: Sustainable Tourism and Ecological Management*. Cambridge: Cambridge University Press.

Amante-Helweg, V. (1996). Ecotourists' beliefs and knowledge about dolphins and the development of cetacean ecotourism. *Aquatic Mammals*, *22*(2), 131–140.

Anderson, L. S., & Heyne, L. A. (2012). Flourishing through leisure: an ecological extension of the leisure and well-being model in therapeutic recreation strengths-based practice. *Therapeutic Recreation Journal, XLVI*(2), 129–152.

Antonioli, C. (2005). *Biophilia: The therapeutic value of animals in the treatment of depression* (PhD), University of Leicester.

Antonioli, C., & Reveley, M. A. (2005). Randomised controlled trial of animal facilitated therapy with dolphins in the treatment of depression. *British Medical Journal*, *331*, 2–4.

Ballantyne, R., Packer, J., & Falk, J. (2011). Visitors' learning for environmental sustainability: testing short- and long-term impacts of wildlife tourism experiences using structural equation modelling. *Tourism Management*, *32*(6), 1243–1252.

Bejder, L., Samuels, A., Whitehead, H., Gales, N., Mann, J., Connor, R., ... Krutzen, M. (2006). Decline in relative abundance of bottlenose dolphins exposed to long-term disturbance. *Conservation Biology*, *20*(6), 1791–1798. doi:10.1111/j.1523-1739.2006.00540.x

Bekoff, M. (2008). Ethics and marine mammals. In W. F. Perrin, J. G. M. Thewissen & B. Wursig (Eds), *Encyclopedia of Marine Mammals* (2nd Edition). Burlington, MA: Academic Press. doi:ProQuest ebrary. Web. 18 February 2016.

Besio, K., Johnston, L., & Longhurst, R. (2008). Sexy beasts and devoted mums: narrating nature through dolphin tourism. *Environment and Planning A*, *40*(5), 1219–1234. doi:10.1068/a38424

Birtles, A., Valentine, P., Curnock, M., Arnold, P., & Dunstan, A. (2002). *Incorporating visitor experiences into ecologically sustainable dwarf minke whale tourism in the northern Great Barrier Reef*. CRC Reef Research Centre Technical Report No 42, CRC Reef Research Centre Ltd, Townsville. Retrieved from: www.researchgate.net

Blewitt, M. (2008). Dolphin-human interactions in Australian waters. In D. Lunney (Ed.), *Too Close for Comfort: Conscientious Issues in Human-Wildlife Encounters* (pp. 197–210). Royal Zoological Society of NSW.

Bowler, D. E., Buyung-Ali, L. M., Knight, T. M., & Pullin, A. S. (2010). A systematic review of evidence for the added benefits to health of exposure to natural environments. *BMC Public Health*, *10*, 456. doi:10.1186/1471-2458-10-456

Bulbeck, C. (2005). *Facing the Wild: Ecotourism, Conservation and Animal Encounters*. Sterling: Earthscan.

Burns, G. L. (2015) Animals as tourism objects: ethically refocusing relationships between tourists and wildlife. In K. Markwell (Ed.), *Animals and Tourism: Understanding Diverse Relationships* (*Aspects of Tourism: Vol. 67*, C. Cooper, Series Ed., pp. 44–59). Bristol: Channel View.

Burns, G. L., MacBeth, J., & Moore, S. (2011). Should dingoes die? Principles for engaging ecocentric ethics in wildlife tourism management. *Journal of Ecotourism*, *10*(3), 179–196. doi:10.1080/14724049.2011.617450

Caldwell, L. L. (2005). Leisure and health: why is leisure therapeutic? *British Journal of Guidance & Counselling*, *33*(1), 7–26. doi:10.1080/03069880412331335939

Caldwell, L. L., & Witt, P. A. (2011). Leisure, recreation, and play from a developmental context. *New Directions for Student Leadership*, *2011*(130), 13–27. doi:10.1002/yd.394

Cloke, P., & Perkins, H. C. (2005). Cetacean performance and tourism in Kaikoura, New Zealand. *Environment and Planning D: Society and Space*, *23*(6), 903–924. doi:10.1068/d57j

Cochrane, A., & Callen, K. (1992). *Dolphins and Their Power to Heal*. London: Bloomsbury.

Constantine, R. (2001). Increased avoidance of swimmers by bottlenose dolphins (Tursiops truncatus) due to long-term exposure to swim-with-dolphin tourism. *Marine Mammal Science*, *17*(4), 689–702.

Constantine, R., Brunton, D. H., & Dennis, T. (2004). Dolphin-watching tour boats change bottlenose dolphin (Tursiops truncatus) behaviour. *Biological Conservation*, *117*(3), 299–307. doi:10.1016/j.biocon.2003.12.009

Curtin, S. (2005). Nature, wild animals and tourism: an experiential view. *Journal of Ecotourism*, *4*(1), 1–15. doi:10.1080/14724040508668434

Curtin, S. (2006). Swimming with dolphins: a phenomenological exploration of tourist recollections. *International Journal of Tourism Research*, *8*(4), 301–315. doi:10.1002/jtr.577

Curtin, S. (2009). Wildlife tourism: the intangible, psychological benefits of human–wildlife encounters. *Current Issues in Tourism, 12*(5–6), 451–474. doi:10.1080/13683500903042857

Curtin, S., & Wilkes, K. (2007). Swimming with captive dolphins: current debates and post-experience dissonance. *International Journal of Tourism Research*, *9*(2), 131–146. doi:10.1002/jtr.599

DeMares, R. (2000). Human peak experience triggered by encounters with cetaceans. *Anthrozoos: A Multidisciplinary Journal of the Interactions of People & Animals*, *13*(2), 89–103.

Dobbs, H. (2000). *The Healing Power of Dolphins*. London: Judy Piatkus.

Duffus, D. A., & Dearden, P. (1990). Non-consumptive wildlife-oriented recreation: a conceptual framework. *Biological Conservation*, *53*, 213–231.

Eckersley, R. (1989). Green politics and the new class: selfishness or virtue? *Political Studies, XXXVII*, 205–223.

Fennell, D. A. (2013a). Contesting the zoo as a setting for ecotourism, and the design of a first principle. *Journal of Ecotourism*, *12*(1), 1–14. doi:10.1080/14724049.2012.737796

Fennell, D. A. (2013b). Ecotourism, animals and ecocentrism: a re-examination of the billfish debate. *Tourism Recreation Research*, *38*(2), 189–202.

Fiedeldey, A. C. (1994). Wild animals ina wilderness setting: an ecosytemic experience? *Anthrozoos: A Multidisciplinary Journal of The Interactions of People & Animals*, *17*(2), 113–123.

Fine, G. A. (1997). Naturework and the taming of the wild: the problem of '"Overpick"in the culture of mushroomers. *Social Problems*, *44*(1), 68–88.

Finn, H., Donaldson, R., & Calver, M. (2008). Feeding Flipper: a case study of a human-dolphin interaction. *Pacific Conservation Biology*, *14*(3), 215–225.

Foroughirad, V., & Mann, J. (2013). Long-term impacts of fish provisioning on the behavior and survival of wild bottlenose dolphins. *Biological Conservation, 160*, 242–249. doi:10.1016/j.biocon.2013.01.001

Frohoff, T. G., & Packard, J. M. (1995). Human interactions wirth free-ranging and captive bottlenose dolphins. *Anthrozoos: A Multidisciplinary Journal of the Interactions of People & Animals, VIII*(1), 44–53.

Garrod, B., & Fennell, D. A. (2004). An analysis of whalewatching codes of conduct. *Annals of Tourism Research, 31*(2), 334–352. doi:10.1016/j.annals.2003.12.003

Gomez-Pompa, A., & Kaus, A. (1992). Taming the wilderness myth. *BioScience, 42*(2), 271–279.

Green, R. J., & Higginbottom, K. (2000). The effects on non-consumtpive wildlife tourism on free-ranging wildlife: a review. *Pacific Conservation Biology, 6*, 183–197.

Haworth, J. T. (1997). *Work, Lesiure and Well-Being*. London: Routledge.

Higham, J. E. S., Bejder, L., & Williams, K. J. H. (2014). Tourism, cetaceans and sustainable development. In J. E. S. Higham, L. Bejder & K. J. H. Williams (Eds), *Whale Watching: Sustainable Tourism and Ecological Management*. Cambridge: Cambridge University Press.

Higham, J. E. S., & Neves, K. (2015). Whales, tourism and mainfold capital fixes. In K. Markwell (Ed.), *Animals and Tourism: Understanding Diverse Relationships* (*Aspects of Tourism: Vol. 67*, C. Cooper, Series Ed.). Bristol: Channel View.

Hill, J., Curtin, S., & Gough, G. (2013). Understanding tourist encounters with nature: a thematic framework. *Tourism Geographies, 16*(1), 68–87. doi:10.1080/14616688.2013.851265

Hills, P., & Argyle, M. (1998). Positive moods derived from leisure and their relationship to happiness and personality. *Personality and Individual Differences, 25*, 525–535.

Hinds, J., & Sparks, P. (2009). Investigating environmental identity, well-being, and meaning. *Ecopsychology, 1*(4), 181–186. doi:10.1089/eco.2009.0026

Hughes, P. (2001). Animals, values and tourism: structural shifts in UK dolphin tourism provision. *Tourism Management, 22*, 321–329.

Iwasaki, Y., & Mannell, R. C. (2010). Hierarchical dimensions of leisure stress coping. *Leisure Sciences, 22*(3), 163–181. doi:10.1080/01490409950121843

Jarvis, C. H. (2000). *If Descartes swam with dolphins: the framing and consumption of marine animals in contemporary australian tourism* (Doctor of Philosophy), University of Melbourne.

Kellert, S. R. (1996). *The Value of Life: Biological Diversity and Human Society*. Washington, DC: Island Press.

Kellert, S. R. (1999). *American Perceptions of Marine Mammals and Their Management*. Washington, DC: Humane Society of the United States.

Kellert, S. R., Gibbs, J. P., & Wohlgenant, T. J. (1995). Canadian perceptions of commercial fisheries management and marine mammal conservation in the Northwest Atlantic Ocean. *Anthrozoos: A Multidisciplinary Journal of The Interactions of People & Animals, 8*(1), 20–30. doi:10.2752/089279395787156518

Kleiber, D. A., Hutchinson, S. L., & Williams, R. (2010). Leisure as a resource in transcending negative life events: self-protection, self-restoration, and personal transformation. *Leisure Sciences, 24*(2), 219–235. doi:10.1080/01490400252900167

Kortenkamp, K. V., & Moore, C. F. (2001). Ecocentrism and anthropocentrism: moral reasoning about ecological commons dilemmas. *Journal of Environmental Psychology, 21*(3), 261–272. doi:10.1006/jevp.2001.0205

Leopold, A. (1989). *A Sand County Almanac and Sketches Here and There*. New York: Oxford University Press.

Lück, M. (2003). Education on marine mammal tours as agent for conservation – but do tourists want to be educated? *Ocean & Coastal Management, 46*(9–10), 943–956. doi:10.1016/s0964-5691(03)00071-1

Lusseau, D., & Higham, J. E. S. (2004). Managing the impacts of dolphin-based tourism through the definition of critical habitats: the case of bottlenose dolphins (Tursiops spp.) in Doubtful Sound, New Zealand. *Tourism Management, 25*(6), 657–667. doi:10.1016/j.tourman.2003.08.012

Lyubomirsky, S., Sheldon, K. M., & Schkade, D. (2005). Pursuing happiness: the architecture of sustainable change. *Review of General Psychology, 9*(2), 111–131. doi:10.1037/1089-2680.9.2.111

Mann, J., Connor, R. C., Barre, L. M., & Heithaus, M. R. (2000). Female reproductive success in bottlenose dolphins (Tursiops sp.): life history, habitat, provisioning, and group-size effects *Behavioral Ecology, 11*(2), 210–219.

Markwell, K. (2015) Birds, beasts and tourists: human-animal relationships in tourism. In K. Markwell (Ed.), *Animals and Tourism: Understanding Diverse Relationships* (*Aspects of Tourism: Vol. 67*, C. Cooper, Series Ed., pp. 1–26). Bristol: Channel View.

Martinez, E., Orams, M. B., & Stockin, K. A. (2011). Swimming with endemic and endangered species: effects of tourism on Hector's dolphins in Akaroa Harbour, New Zealand. *Tourism Review International, 14*, 99–115. doi:10.3727/154427211X13044361606379

Maslow, A. H. (1954). *Motivation and Personality*. New York: Harper.

Mayes, G., Dyer, P., & Richins, H. (2004). Dolphin-human interaction: pro-environmental attitudes, beliefs and intended behaviours and actions of participants in interpretation programs: a pilot study. *Annals of Leisure Research, 7*(1), 34–53. doi:10.1080/11745398.2004.10600938

Mayes, G., & Richins, H. (2009). Dolphin watch tourism: two differing examples of sustainable practices and proenvironmental outcomes. *Tourism in Marine Environments, 5*(2–3), 201–214.

Mehmetoglu, M. (2007). Nature-based tourism: a contrast to everyday life. *Journal of Ecotourism, 6*(2), 111–126. doi:10.2167/joe168.0

Milstein, T. (2008). When whales "speak for themselves": communication as a mediating force in wildlife tourism. *Environmental Communication, 2*(2), 173–192. doi:10.1080/17524030802141745

Milstein, T. (2015). The performer metaphor: "mother nature never gives us the same show twice". *Environmental Communication, 10*(2), 227–248. doi:10.1080/17524032.2015.1018295

Moscardo, G., Woods, B., & Saltzer, R. (2004). The role of interpretation in wildlife tourism. In K. Higginbottom (Ed.), *Wildlife Tourism: Impacts, Management and Planning* (pp. 231–251). Altona, Vic.: Common Ground/ Sustainable Tourism CRC.

Muloin, S. (1998). Wildlife tourism: the psychological benefits of whale watching. *Pacific Tourism Review, 2*, 199–213.

Myers, O. E., & Saunders, C. D. (2002). Animals as links toward developing caring relationships with the natural world. In P. H. Kahn & S. Kellert (Eds), *Children and Nature* (pp. 153–179). Cambridge, MA: MIT press.

Neil, D. T. (2002). Cooperative fishing interactions between Aboriginal Australians and dolphins in eastern Australia. *Anthrozoos: A Multidisciplinary Journal of The Interactions of People & Animals, 15*(1), 3–18.

Neil, D. T., Orams, M. B., & Baglioni, A. (1996). *Effect of previous whale watching experience on participants knowledge of, and response to, whales and whale watching.* Paper presented at the Encounters with whales 1995 Proceedings, Hervey Bay.

O'Connor, S., Campbell, R., Cortez, H., & Knowles, T. (2009). *Whale watching worldwide: tourism numbers, expenditures and expanding economic benefits*, a *special report from the International Fund for Animal Welfare, Yarmouth MA, USA, prepared by Economists at Large.* Retrieved from: www.ifaw.jacksonriverdev.com

O'Neill, F., Barnard, S., & Lee, D. (2004). *Best practice and interpretation in tourist/wildlife encounters: A wild dolphin swim tour example.* Wildlife Tourism Research report series No. 25. Gold Coast, Australia: CRC Sustainable Tourism. Retrieved from: www.sustain.pata.org

Orams, M. B. (1994). Creating effective interpretation for managing interaction between tourists and wildlife. *Australian Journal of Environmental Education, 10*(21), 21–34.

Orams, M. B. (1995). Development and management of a feeding program for wild bottlenose dolphins at Tangalooma, Australia. *Aquatic Mammals, 21*(2), 137–147.

Orams, M. B. (1996). Using interpretation to manage nature-based tourism. *Journal of Sustainable Tourism, 4*(2), 81–94 doi:10.1080/09669589608667260

Orams, M. B. (1997a). The effectiveness of environmental education: can we turn tourists into 'greenies'? *Progress in Tourism and Hospitality Research, 3*, 295–306.

Orams, M. B. (1997b). Historical accounts of human-dolphin interaction and recent developments in wild dolphin based tourism in Australasia. *Tourism Management, 18*(5), 317–326.

Orams, M. B. (2002). Feeding wildlife as a tourism attraction: a review of issues and impacts. *Tourism Management, 23*, 281–293.

Orams, M. B., & Hill, G. (1998). Controlling the ecotourist in a wild dolphin feeding program: is education the answer? *Journal of Environmental Education, 29*(3), 33–38.

Orams, M. B., Hill, G., & Baglioni, A. (1996). "Pushy" behaviour in a wild dolphin feeding program at Tangalooma, Australia. *Marine Mammal Science, 12*(1), 107–117.

Packer, J., & Ballantyne, R. (2012). Comparing captive and non-captive wildlife tourism. *Annals of Tourism Research, 39*(2), 1242–1245.

Passmore, A. (2003). The occupation of leisure: three typologies and their influence on mental health in adolescence. *OTJR, 23*(2), 76–83.

Pergams, O. R., & Zaradic, P. A. (2008). Evidence for a fundamental and pervasive shift away from nature-based recreation. *Proceedings of the National Academy of Sciences USA, 105*(7), 2295–2300. doi:10.1073/pnas.0709893105

Ragheb, M. G. (1996). The search for meaning in leisure pursuits: review, conceptualization and a need for a psychometric development. *Leisure Studies, 15*(4), 245–258. doi:10.1080/026143696375549

Richardson, W. J., & Würsig, B. (1997). Influences of man-made noise and other human actions on cetacean behaviour. *Marine and Freshwater Behaviour and Physiology, 29*(1–4), 183–209. doi:10.1080/10236249709379006

Samuels, A., & Bejder, L. (2004). Chronic interaction between humans and free-ranging bottlenose dolphins near Panama City Beach, Florida, USA. *Journal of Cetacean Research Management, 6*(1), 69–77.

Saunders, R. E., Laing, J., & Weiler, B. (2013). Personal transformation through long-distance walking. In S. Felip & P. Pearce (Eds), *Tourist Experience and Fulfilment: Insights from Positive Psychology* (pp. 127–146). London: Routledge.

Savage, S. (1990). *Dolphins and Whales.* London: Quintet.

Schänzel, H. A., & McIntosh, A. J. (2000). An insight into the personal and emotive context of wildlife viewing at the Penguin Place, Otago Peninsula, New Zealand. *Journal of Sustainable Tourism*, *8*(1), 36–52. doi:10.1080/09669580008667348

Scheer, M. (2010). Review of self-initiated behaviors of free-ranging cetaceans directed towards human swimmers and waders during open water encounters. *Interaction Studies*, *11*(3), 442–466. doi:10.1075/is.11.3.07sch

Seligman, M. E. (2011). *Flourish: A Visionary New Understanding of Happiness and Well-Being*. New York: Free Press.

Servais, V. (2005). Enchanting dolphins: an analysis of human-dolphin encounters. In J. Knight (Ed.), *Animals in Person. Cultural Perspectives on Human-Animal Intimacies* (pp. 323–349). Oxford: Berg.

Smith, H., Samuels, A., & Bradley, S. (2008). Reducing risky interactions between tourists and free-ranging dolphins (Tursiops sp.) in an artificial feeding program at Monkey Mia, Western Australia. *Tourism Management*, *29*(5), 994–1001. doi:10.1016/j.tourman.2008.01.001

Spradlin, T. R., Devenak, A. D., Terbush., & Nitta, E. T. (1999). *Interactions between the public and wild dolphins in the United States: biological concerns and the Marine Mammal Protection Act*. Paper presented at the 13th Biennial Conference on the Biology of Marine Mammals. Maui, Hawaii.

Stamation, K. A. (2008). Too close for comfort: contentious issues in human-wildlife encounters. *Australian Zoologist*, *34*, 211–224.

Steckenreuter, A., Harcourt, R., & Möller, L. (2012). Are speed restriction zones an effective management tool for minimising impacts of boats on dolphins in an Australian marine park? *Marine Policy*, *36*(1), 258–264. doi:10.1016/j.marpol.2011.05.013

Tinsley, H. E. A., & Eldredge, B. D. (1995). Psychological benefits of leisure participation: a taxonomy of activities based on their need-gratifying properties. *Journal of Counseling Psychology*, *42*(2), 123–132.

Tkach, C., & Lyubomirsky, S. (2006). How do people pursue happiness? Relating personality, happiness-increasing strategies, and well-being. *Journal of Happiness Studies*, *7*(2), 183–225. doi:10.1007/s10902-005-4754-1

Turner, J. M. (2002). From woodcraft to 'leave no trace': wilderness, consumerism, and environmentalism in twentieth- century America. *Environmental History*, *7*(3), 462–484.

UNGA (1948). *United Nations Universal Declaration of Human Rights*, United Nations General Assembly.

Unger, L. S., & Kernan, J. B. (1983). On the meaning of lesiure: an investigation of some of the determinants of the subjective experience. *The Journal of Consumer Research*, *9*, 381–392.

UNWTO (1999). *Global Code of Ethics for Tourism*.

Valentine, P. S., Birtles, A., Curnock, M., Arnold, P., & Dunstan, A. (2004). Getting closer to whales – passenger expectations and experiences, and the management of swim with dwarf minke whale interactions in the Great Barrier Reef. *Tourism Management*, *25*(6), 647–655. doi:10.1016/j.tourman.2003.09.001

Wearing, S., & Jobberns, C. (2015) From Free Willy to Seaworld: has ecotourism improved the rights of whales? In C. Cooper (Series Ed.), *Aspects of Tourism. Animals and Tourism: Understanding Diverse Relationships* (pp. 75–90). Bristol: Channel View.

Webb, N. L., & Drummond, P. D. (2001). The effect of swimming with dolphins on human well-being and anxiety. *Anthrozoos: A Multidisciplinary Journal of The Interactions of People & Animals*, *14*(2), 81–85.

White, T. I. (2007). *In Defense of Dolphins: The New Moral Frontier*. Oxford: Blackwell.
Wiener, C. (2013). Friendly or dangerous waters? Understanding dolphin swim tourism encounters. *Annals of Leisure Research, 16*(1), 55–71. doi:10.1080/11745398.2013.768155
Wiener, C. (2015). Dolphin tourism and human perceptions: social considerations to assessing the human-dolphin interface. In K. Markwell (Ed.), *Animals and Tourism: Understanding Diverse Relationships* (*Aspects of Tourism: Vol. 67*, C. Cooper, Series Ed.). Bristol: Channel View.
Wilson, E. O. (1984). *Biophilia, the Human Bond with Other Species*. Cambridge, MA: Harvard University Press.
Zeppel, H. (2008). Education and conservation benefits of marine wildlife tours: developing free-choice learning experiences. *Spring, 39*(3), 3–17.
Zeppel, H. (2009). *Managing swim with wild dolphins tourism in Australia: guidelines, operator practices and research on tourism impacts*. Paper presented at the CAUTHE – 18th International Research Conference 2009, Fremantle, WA, Australia.
Zeppel, H., & Muloin, S. (2008). Conservation benefits of interpretation on marine wildlife tours. *Human Dimensions of Wildlife, 13*(4), 280–294. doi:10.1080/10871200802187105

10 Angler and fish relations in the UK

Ethics, aesthetics and material semiotics

Tom Mordue and Sharon Wilson

Introduction

Freshwater angling is one of the most significant leisure activities in the UK. In 2009, for example, freshwater anglers' gross expenditure in England and Wales was £1.18 billion, generating 37,386 jobs (Radford *et al.*, 2009). Despite such spending power, and the shear popularity of angling (see Mordue, 2009), it is relatively rarely researched, sitting 'uncomfortably' between the different policy and academic foci of sport, leisure, recreation and environment. This chapter goes some way towards redressing that anomaly, and, using a material semiotics approach, situates angling in all these contexts as it examines the social and natural relationships between anglers and fish across the main UK freshwater-angling codes: coarse angling and game angling. From here the question is asked: are wild fish social animals deserving 'sporting justice' or natural beings deserving absolute protection from that human corruption we call angling?

By applying material semiotics we recognise that all angling is made up of a multiplicity of socio-natural practices. These practices include anglers subscribing to certain sporting practices and angling methods, learning about fish species and their habitats, buying and applying equipment in very particular ways, attempting to 'think like a fish' (Bear and Eden, 2011) and becoming part of the fish's environment when angling. The multiplicity of such practices is not only legion in general terms, but varies according to the specificities of the quarry sought and angling methods used. That said, freshwater-angling practices in the UK do aggregate to an overarching bifurcation of angling codes we term coarse angling and game angling, and it is this bifurcation that guides the analysis here.

While coarse and game angling are two sides of the same sporting coin, those two sides are not equal. For example, coarse angling accounts for 87% of all freshwater angling in the UK (Bear and Eden, 2011), and targets essentially any non-salmonid fish that are common to lakes, canals and rivers up and down the UK. While coarse angling has many sub-divisions and specialisms, in general terms, coarse anglers are very catholic in angling methods used, adapting any legal means available to them, with bait-fishing being by far the most dominant. UK game anglers, on the other hand, fish only for trout, salmon and sometimes char, mostly in the rivers where these fish are the predominant species. They are also selective in methods deployed, mostly fishing with lure or fly, with fly-fishing being by far the most dominant method. The differences between coarse and game angling are, however, much more than about differing sporting preferences and practices, they also constitute divisions along social lines. Game angling occupies a more exclusive social position to that of coarse angling – even though its economic significance is much less. Such divisions are also reflected in the way fish are valued, with trout, for example, being conferred a certain 'wily' intelligence and salmon often exulted as 'the king of fish' in the game angling literature (Lapsley, 2003; Osborne, 2007). The coarse-angling literature seems less inclined toward such anthropomorphic levitation, though it is not unheard of for coarse anglers to engage in reactionary retorts by comparing the merits of 'their' fish with those of 'the other side'. For example, in an interview with Kevin Parr for *The Idler* magazine in 2008, Chris Yates, a coarse-angling author celebrated for his literary style, barks:

> I'm not a great fan of salmon and trout … They are pushy sort of fish, always jumping about and being far too flashy and showy. [They're] incredibly thick – because they're like a cruise missile, with the same amount of intelligence, in that they just keep going, no matter what you throw at them, they keep going. Whereas the carp, the carp is the grandmaster of chess.

The absurd humour in the statement serves to illustrate the push back on the self-asserted superiority of game angling, which is obviously disliked in at least some coarse-angling quarters.

Similar imputed superiority extends to game angling's geographies, where the best game-fishing rivers and streams are very exclusive spaces, being either prohibitively expensive to access or simply closed off to the public and the open market altogether. For example, *Country Life* magazine writes of the Houghton Fly Fishing Club on the river Test, which is a river 'world renowned' for its brown trout fishing:

> Membership of the Houghton Fishing Club (HFC) is the dream of every serious fly-fisherman [sic], but is obtainable by only the very fortunate few. Started in 1822, the HFC is the oldest private fishing club in Britain … The club owns – and has exclusive fishing rights on – 13 miles of the River Test, near Stockbridge in Hampshire. Once known as the Stockbridge Fishing Club, the HFC is famously very private, and has a restricted, elected membership of a mere 25. Eminent members [are]: Lord Tanlaw, Lord Tryon and the Duke of Northumberland.
>
> (Country Life, 2008)

Salmon fishing on the river Tweed's 'Junction Pool' near Kelso in the Sottish Borders is similarly famous and exclusive, reportedly costing around £27,000 for a week's hire in October, when the main salmon run happens, to a limit of five anglers who have not only to be rich enough but well-connected enough to access it (Salmon Fishing Forum, 2017). Though such spaces are never physically too far from anyone living in the UK, they are situated way beyond the reach of the vast majority of its citizens who may want to chance their arm. Freshwater fish and the environments where they thrive, then, are as socially embroiled as they are authentically natural.

In consequence, demarcating what are natural or social relations and practices in UK freshwater angling is difficult to fathom. Yet it is at the point of such nature–society cross-sections that this chapter is targeted as it explores the knotted intimacies, relationalities and separations of angler–fish and angler–angler interactions across the two main UK freshwater-angling codes. From this vantage the chapter considers angling and anti-angling interests and antagonisms to answer the question already mooted: are freshwater fish social animals deserving 'sporting justice' or natural beings deserving absolute protection from those human corruptions we call angling?

Fish and angler, nature and society

The material semiotic approach taken here integrates aspects of actor-network theory and 'performativity' (Thrift, 2003) because they complement each other by providing ways of understanding how social and natural forces can come together to create relational effects. This conceptual platform allows a relational focus on the effects of power within and between different angling practices and the actors that *perform* them. From an angling practice perspective, performance is applied to consider how the human world is both socially constructed and embodied, made by natural rhythms and registers as well as cultural representations and discourses. This sits alongside the post-humanism of actor-network theory which sees humans and non-humans as *symmetrical*, in that neither one should be privileged over the other in any a priori sense. Rather, they should be critically evaluated as to how they are able to effect and create the world. Actor-network theory's starting position is that human and non-human actors should only be assessed against what they do and how they exercise power to act and enlist heterogeneous others to build networks of association (Murdoch, 2006). Freshwater angling, then, bifurcated into coarse and game sub-divisions, is a multitude of relational effects made up of heterogeneous human and non-human actors that are assembled together into actor-networks of common endeavour. In this network perspective, fish are co-creators of angling with humans, as are other non-humans like fishing technologies and equipment, rivers and lakes. The issue here is about the varied power anglers and fish have in the angling actor-network and how their interests are shaped and served.

Such human/non-human symmetry does not mean that all actors are equal or should be seen as equal. Rather, that all actors are equally implicated in power relations and some will become more dominant than others. Dominant actors are those who determine action the most, and whether they are human or not is of less importance than their ability to be powerful and enlist others in the actor-network. Murdoch (2006) reminds us, however, that humans do occupy a uniquely powerful place in the world because of their ability to represent and speak for others. The most powerful human actors will therefore gain the capacity to literally speak for other humans and non-humans in a network, giving them enormous potential influence.

The way angling is shaped and performed is thus an effect of particular sets of discourses, material circumstances and embodied practices. These elements also feedback on each other every time an angler engages a fish in the wild and discursively relays his or her experiences to the wider angling network. In this way, successful angling authors gain immense authority in the angling actor-network and are sought out by organisations – such as angling media (magazines and TV, and fishing-tackle manufacturers) – who can profit through their commentaries and their endorsements. As angling communication networks expand, the citational power (cf. Butler, 1993) and impact of these dominant actors also expands and the potential influence of the UK angling actor-network becomes that much greater. Indeed, though modern fly-fishing for trout and salmon was perfected as a unified sport on the rivers of the UK in the late 1800s, through citational performances and other sporting mobilities its codes were soon exported to the USA, and it is now offered as a tourism experience in destinations all around the world (Mordue, 2013).

There is a corollary to such development. When a network establishes and endures over time and expands its reach through space, it will face increasing challenges and perhaps oppositions from other networks and associations, whether they be sporting, cultural, social or moral. For example, the strict UK codes of fly-fishing are often adapted to new local circumstances and environments as the tourism industry offers increasingly exotic destinations to anglers seeking new experiences. Closer to home, the challenge of anti-angling voices in the UK has become louder since environmental and animal rights movements gained momentum in the 1960s, and a power struggle here is always lurking between field sports and environmental and animal welfare interests (more on which later). Moreover, further dispersal of the sport is always a possible threat as different network actors could attempt to assert their own dominance and angling interests over others. In this way they may seek 'lines of flight' (Deleuze and Guttari, 1998) away from the established network to form new angling codes and practices and, therefore, new sub-divisive network arrangements that could make the current bifurcation between coarse and game angling more complex and unstable. Just as with any other successful actor-network, such dynamism and complexity means that network maintenance is a constant requirement in angling if its current sporting configurations and arrangements are to endure and prosper.

Working on the sport and delineating the social

Until the Victorian era, freshwater angling was largely undifferentiated in terms of methods and species sought. Angling was seen as an open and inclusive pastime most famously described and celebrated by Izaak Walton in 1653 as a convivial leisure pursuit that was good for the mind, body and soul; taking men [sic] away from the machinations of everyday life into a world of natural and peaceful simplicity (see Walton, 1962; Franklin, 1999; 2000). On this Paxman (1995: 475) notes:

> Although some Jacobean authors suggested there was a hierarchy in fishing, it took the Victorians, with their fine nose for orders of being, to create the full range of class distinctions. Before then, anglers had been happy enough to use worm, maggot or fly, just as long as they caught a fish or two.

This represented the greatest 'line of flight' (Deleuze and Guttari, 1998) in freshwater angling to-date, which was led by rather privileged men such as F. M. Halford, G. S. Marryat and G.E.M. Skues who themselves privileged fly-fishing for brown trout in southern England chalkstreams above all other means and methods of catching fish.

This development involved considerable actor-networking. For example, Halford, Marryat and Skues each authored what are considered seminal texts on the science and art of fly-fishing in their own rights, but in doing so they vied for leadership and dominance of the sport, purporting certain fly-fishing methods as either superior to or more appropriate than others. Halford was the ultra-purist, advocating a dry-fly-only code where the angler's flies are fished only on the surface of the water, no matter what the ambient weather or water conditions. Marryat and Skues were less dogmatic, and quite oppositionally advocated wet-fly fishing (placing flies below the surface) or dry-fly depending only on ambient circumstances of weather, water and fish. These debates were more than esoteric or simply aesthetic, they constituted 'the most controversial developments the sport [of fly-fishing] has ever seen – indeed, a controversy that would carry right to our own fishing today' (Law, 2017).

Indeed, the substantial, and often very literary, fly-fishing literature of today is still pregnant with these debates, analysing them and retelling them in ways that constantly re-authenticate fly-fishing's 'rightful' place at the zenith of the angling world (see, for example, Lapsley, 2003). These trout fishing men had their salmon fishing counterparts who undertook much of their own fieldwork on the great salmon rivers of Scotland, and who published their findings in books that also became seminal texts. They also imputed Victorian hierarchies and orders of being onto the natures and societies where their quarry was to be had:

> The writings of T. H. White, Frederick Aflalo, Captain Albert F. L. Bacon and countless others record an age of endless salmon and wild moors, peat-fired lodges and wisened ghillies in kilts, and a world of rigid class distinction where everything – including the salmon – had its rightful place.
>
> (Berry 2011, p. 78)

At the same time the landed gentry – particularly those with riparian ownership of the most valued waters – played a major part in designing and ordering a whole new angling licensing regime, as well as the policing of the rivers by bailiffs, and instituting the dominance of fly-fishing as the most appropriate way to catch salmon. As with elite brown trout fishing, salmon fishing cemented its sporting codes and spatial regulations in such a way that it 'rapidly accumulated a burden of tradition and convention which was predominantly elite in character and origin' (Osborne, 2007: 211).

Sporting practices, space and power

As alluded to, coarse angling is much more varied in terms of methods used and species sought than is game angling, and this is arguably a significant reason why it has sub-divided so catholically into quite different specialisms over the last 100 years or so, giving it less coherence than game angling both as a sport and as an angling/social movement. For example, there is match fishing where anglers compete with each other to win cash prizes for the greatest weight of fish taken from a particular stretch of water in a given time. There is also specimen hunting in which anglers target only the very largest of a particular species of fish with the hope of breaking a British record. Many carp anglers, for example, fall into this category. The types of water fished by coarse anglers is also very varied, ranging from rivers, streams and canals to lakes, ponds and gravel pits – many of which can be found in or very close to conurbations. There are also many pleasure anglers who are not too partisan about their fishing and who are happy to catch whichever type of fish appears at the end of their lines wherever they may be fishing. These anglers are more easily placed into the coarse-angling camp than the game-angling camp precisely because of their absence of dogma.

Though, as we have seen, there is some stated distain between game and coarse, these sub-divisions seem to have emerged with little debated controversy within and outwith coarse angling. Perhaps this is because the spaces of coarse angling are less restricted than those of game angling, making room for large populations of coarse fish and myriad methods to catch them. It may also be because coarse angling has not seen the need to be self-consciously ideological in the way game angling has. This is evident from the coarse-angling literature, which is much more instructional, focusing on the 'how to' of practice, rather than on literary treatise on any intertwined sporting etiquette and social status. Indeed, coarse angling has never defined itself, but was given the appellation 'coarse' by a Victorian sporting elite which, in doing so, indicated their disdain for the practice, and no doubt the fish and the people who practised it (cf. Lowerson, 1993).

Having said all this, it would be wrong to imply that game angling is only an upper-class activity and that coarse angling is left to those who cannot cross the requisite social threshold to practise it. The boundaries between the two are not so clear cut but they are, nonetheless, there. For example, game angling has become quite widely practised in the UK in recent decades as market forces have sought to profit from it by growing the market and widening access to it. In this, there has been: mass production of instructional texts and equipment; commonplace stocking of rivers with brown trout and salmon; de-industrialisation, which allowed the cleaning up of once polluted rivers with game-angling potential; and by creating purpose built game-angling lakes close to conurbations. While these developments have no doubt widened access, they are anything but threats to the hallowed rivers, streams, clubs, traditions and dominant actors that make up and sustain the game-angling sporting tabernacle. Indeed, the provision of popular game angling has served to protect such secluded waters by diverting the popular angling gaze away from them and onto spaces where the elites of the sport have little desire to tread or colonise. Moreover, the sophistication of the game-angling literature, with its emphasis on sporting codes and traditions as well as instruction, ensures that game angling is as tightly and 'authentically' practised and protected as it ever was.

All this social work not only reveals a particular ideological framing of nature and society but implies a certain deliberation and assertion of power in which game angling simply out-muscles coarse angling in social, if not financial, terms. However, and notwithstanding all this, it would be wrong to think that such imposition is only a matter of force applied from the top of society down. Things are more subtle than this. The way power operates in angling involves nuanced as well as more obvious interchanges between material and semiotic forces, and happens in a Foucaultian sense whereby power and knowledge are indissociable and mutually reinforcing. Power/knowledge is thus integral to concrete practices (Law, 1986; Fox, 2000) whereby certain knowledges reproduce certain power relations that produce effective actions in (and from) particularly effective spaces. For instance, Murdoch (2006) tells us that in early actor-network studies laboratories were the focus of analysis because they are specialised spaces that produce scientific knowledge that may become effective. That is if knowledge born in the laboratory is successfully circulated in wider society to influence everyday practices, then it assumes power with effectivity that is way above and beyond its value to science. Scientific knowledge, then, does not have intrinsic power of itself, rather power gathers through its interpellation and translation into everyday life, and that takes actor-networking. Such power can be either a force for good or ill and, in Foucault's words, comes from effective locales and travels through 'the moving substrata of force relations which, by virtue of their inequality, constantly engender states of power' (Foucault, 1984: 93). Therefore, to create and sustain a cultural field such as game angling and its opposite, coarse angling, power needs to be harnessed from effective local actions and spaces, such as the laboratory work in the chalkstreams of England and/or the salmon rivers of Scotland, and through discursive techniques, such as publishing empirical findings, directed toward the purpose of network building in wider society. However, once built, those networks are never completely stable, and are always liable to external and internal threats. It is thus the ability of an actor-network to hold together in the face of such dynamism that is the most important attribute if an actor-network is to endure (Murdoch, 2006). This is why there needs to be so much vigilance exercised in maintaining the boundaries and hierarchies in angling, not least those separating game and coarse angling, if current arrangements are to be maintained.

Sporting aesthetics and ethics in game and coarse angling

It is clear, then, that angling is socially constructed and is divided by bifurcated power struggles and performances that are reflective of power geometries in wider society. That is not to say that the coarse- and game-angling apartheid simply reflects or represents wider social divisions, it is to say that they are both implicated in and constitutive of, as well as outcomes of, wider power struggles. As such, they are material semiotic regimes that link the natural world and the social world in very particular, dialectical, ways and operate through clearly bounded networks that produce quite distinct sporting outcomes both for anglers and for fish. For this reason, it is worth examining the aesthetics and ethics of these sporting codes a little more and considering whether these different codes of practice produce comparatively more or less ethical angler–fish relations.

As we have noted, fly-fishing is at the very heart of game angling, putting tight restrictions on how one might catch a trout or a salmon most sportingly, with all the best trout streams and salmon rivers insisting on a 'fly-only' code, and with some trout streams even restricting anglers to 'dry-fly only'. Similarly, the game-angling literature lauds fly-fishing as an authentic and highly skilful craft that is intimately in touch with nature because it uses as light and little tackle as possible, and is therefore unencumbered by gismos and suchlike that would interfere with that authenticity. On the other hand, coarse angling is less precious about such authenticity. Indeed, coarse angling embraces technology, both in terms of equipment and baits, which allows the advancement of any legal practice that helps catch fish. These different stances toward craft and technology underscore very fundamental differences in sporting values, sporting practice and ethics of 'fair play' between the codes. And, as Hummel (1994) points out, a fair play ethic is the capstone to any sport-fishing practice. Table 10.1 is an attempt to summarise these differences and illustrates that coarse angling is comparatively democratic because of its inclusivity, mass appeal and accessibility.

Fly-fishing, on the other hand, is ethically elitist because it is selective, takes time and practice to learn, even at a rudimentary level, and is cerebral, mostly eschewing bait-fishing in favour of fishing with hand-made flies that mimic natural insects. Indeed, fly-fishers often lambast coarse angling as simply feeding fish with hooked bait, whereas fly-fishing is about out-foxing fish on their terms by ingeniously presenting a hand-made lure that is apparently natural to them. Fly-fishing

Table 10.1 Comparison of ideal–typical forms of angling in the UK

	Coarse angling	*Game/fly-fishing*
Value orientations	Democratic/inclusive	Elitists/exclusive
Goals	Most/biggest fish	Most difficult fish
Means	Technology (latest gadgets)	Craftsmanship (lightest tackle)
Standards of performance	Results of performance (size of catch)	Quality of performance (how caught)
Rewards	External (displayed skills, public esteem, trophies)	Internal (self-satisfaction)
Participants	Mass appeal	Selective appeal
Technology	Promoted	Resisted

Source: Adapted from Hummel (1994: 45).

is also predominantly about the individual angler angling for personal pleasure and personal sporting advancement in a natural environment that allows unencumbered individual wandering to hunt and search for catching opportunities. Indeed, on the best game-angling rivers the numbers of anglers on any one 'beat', or stretch of water, is highly restricted. Coarse angling is much more gregarious and static as anglers bait particular swims to attract fish towards them. It can also be all about competition between the anglers themselves, as in match-fishing for example, which is the most instrumental and Taylorist form of fishing there is. While fly-fishing does sometimes hold competitions, it is not characterised by them in the way coarse angling is. The norm in game angling is to pitch one's wits only against the quarry, not one's fellow anglers. Moreover, 'catch-and-release' fishing has always been the preserve of coarse angling, apparently because they are not good eating, while game anglers traditionally killed and kept their more tasty catch 'for the pot'.

In recent years, however, catch-and-release has become common practice in game angling, particularly at key times of the year when, for example, salmon runs or trout populations are not at their strongest. Thus by returning caught fish to the river the maximum number of fish can be encouraged to spawn and stocks are replenished naturally. When on those occasions catch-and-release policies are lifted – usually because of the plenitude of healthy fish at a given time of year – restrictions on the number of fish an angler can take are still often applied. This is not simply about managing populations of game fish in rivers, it is about maintaining and enhancing the authenticity of those stocks by minimising the need to compensate for the loss of wild fish by restocking with farmed fish. 'Truly' wild trout and salmon are the greatest prize in game angling precisely because they are seen by game anglers as untainted by human hand and are therefore the most natural and authentic of fish to be had in British rivers. Moreover, a crucial aspect to this authenticity is that authentically wild fish have provenance in that they are of that particular stream or river, and the more a river produces and reproduces fish authentically the more valuable that river will be both in cultural and economic terms. Indeed, the best of these rivers, and the best parts of them, are largely unavailable to the market or if they are they command such high rental fees that they prohibit all but the very well healed and the very well connected. Thus. while practices of managing and protecting fish populations in UK freshwaters are becoming more convergent, the sporting aesthetics of catching fish in game and coarse angling remain as bifurcated as ever.

Is game more ethical than coarse? To answer this question it is worth consulting a long-standing treatise on the ethics of different angling practices as voiced in an influential essay by A. A. Luce, a Trinity College, Dublin, Professor of Moral Philosophy in the 1950s, and game-angling man [sic]. Luce argues that catch-and-release is wrong and unethical because sport alone does not justify even the admittedly 'mild form of cruelty' (1959: 179) involved. 'The primary object of justifiable angling is to catch fish for food; there are various pleasures incidental to angling; but they cannot justify the infliction of pain or death' (ibid., 180).

When Luce was writing, catch-and-release in game angling was not the norm, and killing fish for the pot was. Luce's beloved game angling therefore survives his ethical terms while coarse angling does not. Whatever one may think of the ethical rights and wrongs of such an argument, for Luce's ethic to work in practice, while maintaining the sport's authenticity, the population of regular game anglers would have to be correspondingly small relative to the population of fish to sustain wild fish stocks. This of course means that game angling is ethical and authentic as long as it is kept relatively exclusive.

The sustainably modern and more democratic answer to this conundrum is catch-and-release, which brings game angling closer to coarse angling than ever in terms of the practicalities of fisheries management. Other sustainable activities such as conserving waterways, the wildlife, environments and the ecologies that fish and anglers depend upon and enjoy are also integral to modern practices of all freshwater angling. There are also degrees of cruelty in both types of angling, but just because there is cruelty in both it does not mean that all anglers do not care for fish welfare. Some anglers will treat individual fish with more care than others of course, but in general terms anglers do care about the welfare of fish. Therefore, given the wide practice of catch-and-release, the environmental and ecological management practices by anglers in general, and the levels of cruelty across both codes, it is difficult to argue that one is any more ethical than the other.

It is also true that underpinning the care for fish that anglers in general exhibit is a deeply held passion for the sport of catching them. This passion, care, cruelty triad seems contradictory, but it is nonetheless real. The next section focuses on this contradiction as it considers the universal passion for angling that anglers share against those non-anglers who share a passion for banning it.

Tangled ethics and corruptions: a passion for and against angling

All angling literature, whether it be coarse or game, proclaims a deeply held passion for the sport that drives anglers to leave their beds on cold mornings, to travel significant distances and/or pay significant sums just to put a line in the water. Equally, there is the anti-angling lobby who see all angling as both cruel and pointless in this day and age. We discuss these passions here because they are significant both to the development and the future of this ever popular and pervasive of modern pastimes.

A review of the book *A Passion for Angling* (1993), which accompanied the BBC TV series of the same name, sheds some light on what draws men [sic] to fish. Both book and series crossed all freshwater-angling boundaries, and are therefore quite even-handed in their treatment and promotion of freshwater angling. This is what the review has to say:

> The book, and the film series, sets out to capture the very essence of fishing. That almost intangible thing that draws grown men to the river bank, in the manner and wonder of a small boy, every weekend. That undefineable,

> certain something that we as anglers all know, but can never put our fingers on. The book does not elucidate what this something is, to do so would be to destroy the magic, but it does show the beauty and wonder remains for years and years. Ephemeral, and elusive like many of the quarry species, but real and vibrant nonetheless.
>
> If you are an angler, or you require to develop an understanding of a husband, boyfriend, or brother that fishes you could do worse than read this book. The answer is not there, but it will give you a measure of understanding as to what drives them. Maybe, the only way to really understand is to grip the rod yourself and follow the dream. To adopt and embrace the Passion for Angling.
>
> (Amazon customer, 2003)

Notwithstanding the gender bias in this review, it is reflective of angling being a largely male sport, and, ending as it does on an emphasis on doing rather than discoursing, it stresses the physicality, the embodied nature and the non-representational mysteries of angling practice. In other words, it alludes to how angling is a performed relational effect in which body and spirit along with culture and society enmesh when an angler takes to the water to catch and bring his quarry from its watery world into the airy world of society. After Deleuze and Guttari (1998), Bear and Eden (2011) say that in doing this the angler comes to 'think like a fish' and in that quest 'become-fish' as an essential part of learning to become-angler. A passion for angling then is about loving nature in the form of fish and fishy environments, but it is also about domination of those.

Because being a successful angler means being able to get close to fish in their natural state in these ways, Franklin (2001) argues that angling is a much more profound form of leisure and ecotourism than any other. Moreover, Bauer and Herr (2004) assert that angling, as with hunting, is an elemental practice appealing to our 'killer ape' instincts. It is, therefore, a doing that can only be partially revealed by social constructionist accounts that emphasise the gender, culture and class-based divides that structure what it is and how it is performed. Thus, while angling has been increasingly shaped by culture it cannot be denied by culture because it sits at the very crossroads of society and nature, human and animal, emotion and instinct. And it is in this material semiotic reciprocation where a universal passion for angling resides, and where the power of dominant angling actors is focused as they connect the ways of the natural to divisions of the social, creating the bifurcations and unions between and within coarse and game angling discussed above.

Ironically, then, passion is a unifying force in a sport that is anything but united, and, as alluded to, is the prime motivator that compels the angler to get to the water's edge in a prepared state of readiness for 'the catch'. Catching a fish, then, can be something of a climactic event which instantaneously releases days, weeks and even years of anticipation and learning. In a discussion of fly-fishing in these terms, Preston-Whyte (2008: 53) describes the moment of the catch as:

> interweaving [a] sense of abandonment to an external force at the moment the fish strikes and during the fight for its life, with the delight and satisfaction afforded by casting virtuosity, hunting skills, and local environmental fish lore … Deeply buried primeval urges surface that glory in the lust for the hunt. Time comes to a standstill. The fisher inhabits a liminal space between the moments before the fish takes the hook until after its successful capture.

Couple this elementally visceral moment with a passion and appreciation for simply being in the environments where freshwater angling takes place, then a heady mix of dualities – such as genetics and aesthetics, restraint and lust, cruelty and welfare, town and country, culture and nature, regulation and escape, and love and domination – can be seen to be co-presences in the world of angling. Moreover, if we examine just one of these co-presences, love and domination, and pick apart the contradictions it contains, something of a relational window begins to open through which we can view and situate many of the arguments for why angling is such an importantly compulsive pastime and those against angling as a pointlessly cruel activity.

First off, from the time of Izaak Walton to the present day, the angling literature expresses universal love not only for the quarry but for immersion in the environments in which fish are found. Predictably, it is also true of the literature, especially the game-fishing literature, that the wilder the fish and the wilder the environment the greater the love expressed seems to be. In recent times, however, the expression of this love has taken an interesting turn. For example, Washabaugh and Washabaugh (2000) notice that touching and loving fish in the USA has become more demonstrable both in USA fly-fishing TV programmes and in first-hand observation of fly-fishing practice where anglers would commonly kiss fish at the time of capture. In the UK, angling TV programmes now also commonly show game and coarse anglers kissing their prize before its release back to the watery world from whence it came. This is not to say that this is an essential or universal practice either in the UK or in the USA, but it is apparently common enough that we may ask: what can we read into this growing penchant for kissing fish?

It appears that the kiss is an anthropomorphised expression not only of love for the wild creature but an act of grace and favour as the fish is permitted to leave the airy world of the angler to return to its natural state. If the angler was starving, this of course would not happen. Love and domination are thus meted out simultaneously in a single, sporting act even if the love/domination dualism is not obvious in the way the performance is usually enacted and perceived.

> Domination, in the stream as well as in social life beyond it, is normally cloaked, guarded, and hidden from view. One sees the kiss, the adulation, the celebration; but in trout fishing one rarely sees the sword and the thrust. Especially where fly-fishing is pursued with a catch-kiss-release style … [o]ne can recognize the affections with no difficulty, but not the domination.
>
> (Washabaugh and Washabaught, 2000:129–130)

Back in society, domination is invariably hidden from view in representations of angling by anglers. For example, on UK TV when a game fish is sometimes (still) despatched for the pot, that moment is hardly, if ever, shown.

However, what is clear to the keen observer is that, whether fish are kissed or not, all freshwater angling involves hooking fish and bringing them out of their watery worlds, which must be stressful for the individual fish and therefore cruel. Moreover, it is done these days overwhelmingly for sport and not for sustenance. Should we not then refrain from kissing fish altogether and leave them to swim unhindered by human hand or mouth in their natural state of being? Following Luce's (1959) argument, modern angling in the UK is largely playing with wild creatures and is therefore unethical, which is an argument that the most tolerant in the anti-angling lobby would no doubt subscribe to – the more militant, of course, would ban any sort of angling, including fishing 'for the pot'. However, while there is some congruence between ethical positions in the anti-angling lobby with that of some game anglers, there are also similar contradictions in the anti-angling passion for banning or curtailing angling that are open to criticism. For example, by applying Singer's (1990) animal liberation philosophy the anti-angling lobby simply assign a sentimental anthropomorphism to fish in claims that fish feel pain and anxiety in similar ways to humans would if literally dragged from their lifeworld. There is, however, no concrete scientific evidence to support this claim (Rose, 2007; Newby and Stevens, 2008; Arlinghaus *et al.*, 2009), though equally there is no scientific evidence that categorically denies it (Arlinghaus *et al.*, 2009; also Arlinghaus *et al.*, 2012). This said, we must not fall into the dualistic trap of arbitrating moral positions on positivist scientific grounds when those grounds are incomplete and have their own existential and philosophical limitations. It is useful, however, to apply these different subjective/objective positions to illustrate how each side of the dualism can deny the 'other', and exercise a will to overpower it. It also helps to clarify that it is in the space that lies between, across and outside these two positions that our material semiotic position straddles, connects and challenges via its networked relational ontology.

As we have seen, freshwater angling is an actor-network with a long heritage that is part of the very fabric of British society. It has an extensive network of clubs and affiliations, it is publicly accepted, and it has a government agency, the Environment Agency, which regulates, promotes and supports angling as a wholesome practice deserving continued public support. As such, the anti-angling lobby has a significant and complex regime to overturn if it is to get its way on banning angling on the grounds of animal cruelty. To put it bluntly, it is difficult to see the anti-angling lobby gaining enough power to stop freshwater angling in the UK. That is not to say it is devoid of power, or indeed that it does not have an important cause, but that its real power lies more in its potential to influence angling practice with regard to the welfare of fish, rather than to overthrow angling altogether. Angling will probably always involve some level of cruelty, and the level of cruelty may always be impossible to determine, but there is much common interest between anglers and anti-anglers with regard to either minimising cruelty or ridding angling of cruelty altogether. It is in this shared concern that mutuality exists, but that mutuality will always sit alongside the many socio-natural contradictions and complexities that divide the human actors involved as each side debates the rightfulness of their positions, while the public, by and large, are unaffected and unconcerned.

Conclusion

To return to our question: are wild fish social animals deserving 'sporting justice' or natural beings deserving absolute protection from that human corruption we call angling? As one might expect, the answer is not a simple one, but given the relationalities involved in angling, it is clear that fish are natural beings who contribute enormously to society not only as a food source but as a means of sporting pleasure. The animal liberation position casts them differently as suffering victims belonging only inside nature, away from human harm. Likewise, anglers are cast as either being outside of nature or that they should be separated from it because they value fish only as playthings, and at worst things for cruel play. In this schema, elite game angling, *à la* Luce (1959), has a chance of survival but coarse angling has none, and 87% of angling at the very least would cease. The social impact of this would be significant and regressive, but is that a fair price to pay for the protection of fish?

On this, lessons can be learned from Germany and Switzerland where the Luce (1959) ethic has won out and catch-and-release angling has been banned, though fishing for the pot has not. In both countries there has been a counterintuitive decline in total fish welfare as more fish are unnecessarily killed and as the political clout of recreational anglers, and the sustainable management of fish stocks by them, has been seriously diminished (Arlinghaus *et al.*, 2009, 2017). This is not to suggest that only anglers should be the guardians of fish interests, though it is recognition that recreational angling is a sustainable, albeit socially unequal, practice in the UK. It is also important to recognise that although fish are actors with social impact they cannot speak for themselves, so natural justice is denied them. Instead, as happens with non-humans in all actor-networks (cf. Murdoch, 2006), fish have to rely on humans to speak on their behalf. This means that fish are caught up within socially unequal angling performativities as well as animal welfare discourses, though it is also true to say that fish welfare is important to anglers and non-anglers for numerous compassionate and sporting reasons. Is it not apt and realistic, then, to acknowledge fish within these tangled socio-natural relationalities as actors possessing intrinsic value while having rights to social justice?

Such a relational perspective could override entrenched positions which see fish as either innocents needing protection from certain humans, or as simply food, or as metaphors for naturalising unequal social relations, or as objects of scientific evaluations justifying their status on whether or not they feel and fear like humans. Indeed, viewing fish as material semiotic agents within the angling actor-network positions them socially as well as naturally and repositions the animal welfare arguments aimed at protecting them both inside and outside that network. In this way, social justice sits with sustainable angling practice and eschews socially divisive aesthetics masquerading as sporting ethics. Moreover, the oppositional binaries of fish and angler, nature and society, sentiment and objectivity, angler and anti-angling are reset as dialectical relations that are neither purely good nor bad or amenable to a 'solution' one way or the other. Instead, resolution is a continued and imperfect process where justice for fish and the angling public is not about making them harmonious bedfellows or separating them as natural/social antagonists or indeed casting some anglers as more worthy than others. Rather, it is about resetting the debate in which it is acknowledged that angling produces all sorts of tangled relational effects that not only raise legitimate moral and scientific concerns and interests but ones that have very real socio-natural and political implications.

References

Amazon Customer (2003) Review of C. Yates, *A Passion for Angling* (1993). www.amazon.co.uk/Passion-Angling-Christopher-Yates/dp/0563367415

Arlinghaus, R., Alós, J., Beardmore, B., Daedlow, K., Dorow, M,. Fujitani, M., Hühn, D., Haider, W., Hunt, L. M., Johnson, B. M., Johnston, F., Klefoth, Y., Matsumura, S., Monk, C., Pagel, T., Post, J. R., Rapp, T., Riepe, C., Ward, H., and Wolter, C. (2017) Understanding and managing freshwater recreational fisheries as complex adaptive social-ecological systems. *Reviews in Fisheries Science and Aquaculture* 25, pp. 1–41. Published online October 2016.

Arlinghaus R., Schwab A., Cooke S. J., and Cowx, G. (2009) Review paper: Contrasting pragmatic and suffering-centred approaches to fish welfare in recreational angling. *Journal of Biology* 75, pp. 2448–2463.

Arlinghaus, R., Schwab, A., Riepe, C., and Teel, T. (2012) A primer on anti-angling philosophy and its relevance for recreational fisheries in urbanized societies. *Fisheries* 37 (4), pp. 153–164.

Bauer, J. and Herr, A. (2004) *'Hunting and Fishing Tourism' in Wildlife Tourism: Impacts, Management and Planning*. Altona, Australia: Common Ground Publishing, pp. 57– 78.

Bear, C. and Eden, S. (2011) Thinking like a fish? Engaging with nonhuman difference through recreational angling. *Environment and Planning D: Society and Space* 29, pp. 336–352.

Berry, J. (2011) *A Train to Catch: A Return Ticket to the Golden Age of Fishing*. Ellesmere, Shropshire: Medlar Press.

Butler, J. (1990) *Gender Trouble: Feminism and the Subversion of Identity*, New York: Routledge.

Country Life (2008) Clubs you cannot join. www.countrylife.co.uk/country-pursuits/clubs-you-cannot-join [accessed February 2017]

Deleuze, G. and Guattari, F. (1998) *A Thousand Plateaus*. London: Continuum.

Foucault, M. (1984) *The History of Sexuality*. Volume 1: An Introduction. R. Hurley (Trans.). Harmondsworth: Penguin.

Fox, S. (2000) Communities of practice, Foucault and actor-network theory. *Journal of Management Studies* 37 (6), pp. 853–868.

Franklin, A. (1999) *Animals and Modern Cultures: A Sociology of Human-Animal Relations in Modernity*. London: Sage.

Franklin, A. (2001) Neo-Darwinian leisure, the body and nature: hunting and angling in modernity. *Body and Society* 7 (4), pp. 57–76.

Hummel, R. (1994) *Hunting and Fishing for Sport: Commerce, Controversy, Popular Culture*. Bowling Green, OH: Bowling Green State University Popular Press.

Lapsley, P. (2003) *River Fly-Fishing: The Complete Guide*. London: Robert Hale.

Law, G. (2017) Halford and Skues: "This chalkstream ain't big enough for the both of us". http://midcurrent.com/history/halford-and-skues-this-chalkstream-aint-big-enough-for-the-both-of-us/ [accessed February 2017].

Law, J. (1986) Editor's introduction: Power/knowledge and the dissolution of the sociology of knowledge. In J. Law (Ed.) *Power, Action and Belief: A New Sociology of Knowledge*. London: Routledge and Kegan Paul.

Lowerson, J. (1993) *Sport and the English Middle Classes 1870–1914*. Manchester: Manchester University Press.

Luce, A. (1959) *Fishing and Thinking*. Shrewsbury: Swan Hill Press.

Mordue, T. (2009) Angling in modernity: A tour through society, nature and embodied passion. *Current Issues in Tourism* 12 (6).

Mordue, T. (2013) Travels in to nature and society with rod and line. *Annals of Tourism Research* 43, pp. 100–120.

Murdoch, J. (2006) *Poststructuralist Geography*. London: Sage.

Newby, N. C. and Stevens, E. D. (2008). The effects of the acetic acid "pain" test on feeding, swimming, and respiratory responses of rainbow trout (*Oncorhynchus mykiss*). *Applied Animal Behaviour Science* 114, pp. 260–269.

Osborne, H. (2007) The development of salmon angling in the nineteenth century. In R.W. Hoyle (Ed.) *Our Hunting Fathers: Field Sports in England after 1850*. Lancaster: Carnegie Publishing, pp. 187–211.

Parr, K. (2014) Conversation with Chris Yates. *Idler Magazine*. https://kevinparr.net/2014/11/23/the-idler-magazine-conversation-with-chris-yates/

Paxman, J. (1995) *Fish, Fishing and the Meaning of Life*. Harmondsworth: Penguin.

Preston-Whyte, R. (2008) The lure of fly-fishing. In B. Lovelock (Ed.) *Tourism and the Consumption of Wildlife: Hunting, Shooting and Sport Fishing*. London: Routledge, pp. 45–55.

Radford, A., Riddington, G. and Gibson, H. (2009) The economic evaluation of inland fisheries: The economic impact of freshwater angling in England and Wales. *Environment Agency Science Report – SC050026/SR2*.

Rose, J. D. (2007). Anthropomorphism and the 'mental welfare' of fishes. *Diseases of Aquatic Organisms* 75, pp. 139–154.

Salmon Fishing Forum (2017) www.salmonfishingforum.com [accessed February 2017].

Singer, P. (1990) *Animal Liberation*. New York: Avon Books.

Thrift, N. (2003) Performance and .... *Environment and Planning A* 35, pp. 2019–2024.

Walton, I. (1962) *The Compleat Angler*. London: J.M. Dent & Sons.

Washabaugh, W. and Washabaugh, C. (2000) *Deep Trout: Angling in Popular Culture*. Oxford: Berg.

11 Do wild canids kill for fun?

Robert G. Appleby and Bradley P. Smith

Introduction

On occasion, after a successful kill, predators do not always consume every edible portion of a carcass. The phenomenon of not eating all that is captured is often referred to as excess or surplus killing. Kruuk (1972) defines surplus killing as the killing of prey by a predator, without the killing individual, or members of its social unit, consuming any part of the kill. This is despite the particular prey species being ordinarily eaten by that predator, and free access to the carcass. Some argue against the use of the term surplus killing altogether because of the negative connotations associated with it, preferring instead the term partial prey consumption, as this is so often a major feature of such events (Vucetich *et al.* 2012). Rather than being aberrant, surplus killing is not uncommon in the natural world – it has been observed in a wide range of species, including zooplankton, spiders, predaceous mites, insects, shrews, weasels, marsupials, canids, bears and humans (Vucetich *et al.* 2012). There are many questions about its function, if any, and the causes behind it. Such questions may have important implications in our views of predators, and as we will see, they are more complex than they might initially seem.

Why surplus killing?

In a book about leisure, it might seem strange to explore the rather macabre subject of surplus killing. However, we chose it expressly because predators killing in excess of their immediate need is often used to argue that they are killing for sport, or fun – and thus, could be considered a leisure activity. Such viewpoints are particularly apparent for members of the Family Canidae, who are often seen as 'killing machines', capable of just about killing anything they want (Mech *et al.* 2015).

Adversarial viewpoints about wild canids tend to emphasise their role in livestock losses, and surplus killing can evoke particularly emotional responses in this regard. Understandably, the loss, or the fear of losing livestock to canids is arguably one of the most pertinent aspects of landholder–carnivore conflict around the world (Treves and Karanth 2003). To highlight such landholder perspectives on surplus killing, we use the case of the dingo, a free-ranging Australian canid with an ongoing conflict with landholders (Smith 2015). The theme of dingoes killing for 'sport' and 'for fun' has echoed in the opinions of livestock producers since the European settlement of Australia, and continues to this day. Helen Cathles, a grazier running sheep and goats in the state of New South Wales, summarised some Australian landholders perspective on dingoes quite well when she stated that

> there are many, many ways livestock can die at the whim of a dingo or a wild dog. Food and sustenance are not the only reasons for attacks and killings because, like us, dingoes and wild dogs like to play. Some just kill for the fun of it.
>
> (Cathles 2001, 76)

And from an article more than 50 years earlier, Leo Doyle, an experienced wild dog trapper, reflects that "after satisfying their hunger, it is their habit to kill for sport; hence a couple of dingoes will kill a number of sheep in one night" (*Cootamundra Herald*, Friday 27 August, 1948, 2). Note that the act of killing assumed to satiate hunger is viewed differently from killing where hunger or satiation are not assumed. In other words, it seems apparent that it is not necessarily the loss of one or two sheep infrequently that is the major concern, but the potential to lose many that is at the heart of much of the conflict. This view is supported by W.P. Bluett, who said that:

> It is not the sheep he kills for his 2-day meal that worries the grazier, so much as those he kills for the fun of it. The dog will come on a sheep camp in the early dawn and before the muddled sheep know, he is among them, he has killed half-a-dozen.
>
> (*The Canberra Times*, Wednesday 18 July, 1956, 2)

This is an obvious source of frustration for landholders, who are losing livestock with seemingly no justification on behalf of the animal 'responsible'. And it is a long-running frustration, with John Thistleton repeating a similar sentiment more recently: "They chase the sheep for fun …. They like the live kill and they eat the kidneys. They eat the best parts of the sheep and they don't need to eat [poison] baits" (*The Canberra Times*, Monday 15 July, 2013).

It is not surprising that farmers who, upon discovering that a wild dog has killed, yet not eaten, a number of their livestock in one night, believes that rather than such an event resulting from necessity, it instead stemmed from enjoyment. A predator appearing to kill more than it can immediately consume does not seem to fit with the more usual and serious business of survival, which in a sense, might be forgiven. And it is not just farmers who are perplexed. After all, hunting and killing prey animals can sometimes be very dangerous, and taking apparently unnecessary risks to kill more than is needed to survive and successfully reproduce seems at odds with our understanding of how wild animals should operate.

Why risk injury or worse, hunting and killing prey if not to then eat it? Renowned zoologist Hans Kruuk (1972) raised, amongst others, the question of whether surplus killing constituted wastefulness, which is an obvious question to ask. He concluded that, on face value, it probably was wasteful. Yet, as we will see, Kruuk (1972) and several other eminent researchers have also found some compelling reasons for surplus killing to take place.

Misinterpretation of their behaviour and the motivations behind it has resulted in, and perpetuates, a bad reputation for wild canids (and arguably many predators; Vucetich *et al.* 2012). Surplus killing events only seem to make this worse. But a closer look at the occurrence of, and factors relating to, surplus killing, shows that it is a complicated and multifaceted problem. In order to address the question of whether surplus killing can be considered 'fun' (or indeed constitutes leisure), we explore what we consider to be the key ultimate (evolutionary) and proximate (behavioural, cognitive and neurological) perspectives on the matter. To do this, we address six key areas:

1 Optimal foraging theory and partial prey consumption
2 The vulnerability of prey (both wild and domestic)
3 Fixed action patterns and the ethology of hunting
4 Cognitive explanations in relation to hunting
5 Neurological reward mechanisms
6 The relationship between hunting and play

Evolutionary explanations

Evolution is often considered the 'ultimate' cause of behaviour, because behaviour (or any other trait) that increases survival or fitness of an animal will be selected for. In predator–prey systems, it is a useful analogy to think of an escalating arms race arising, whereby, for example, an advantageous adaptation on the part of the predator may change the selection pressure on the prey (Dawkins and Krebs 1979). Therefore, prey better able to avoid or outmanoeuvre predators will be selected for, at least until the next predator adaptation, and the cycle continues. In truth, interspecific interactions are more complex than just this one possibility, and arms races may stabilise or diminish over time (Dawkins and Krebs 1979), but it helps to illustrate the overarching role that evolution plays in shaping these interactions. Exploring some potential evolutionary explanations for surplus killing therefore seems a fitting starting point.

Kruuk (1972) was perhaps the first biologist to produce a synthesis on the surplus killing phenomenon. He discussed what he saw as the advantages and disadvantages of surplus killing on the part of predators. The ability to return to the kill later, the ability to feed offspring or members of a social group and gain experience in hunting and killing were highlighted as potential advantages. All three may have evolutionary underpinnings in relation to increased survival and fitness. As we previously mentioned, the main disadvantage that Kruuk (1972) identified was the apparent wastefulness of killing prey that were not consumed, either immediately, or later, by the predator or its social group members. However, from an evolutionary standpoint, any apparent wastage would have to affect survival or fitness of the predator in order for such behaviour to be selected against. Given the rarity of most surplus killing events, such a selection force is usually likely to be minimal or non-existent (Miller *et al.* 1985). Building on Kruuk's (1972) seminal exploration, we will delve further into some of the possible evolutionary roots of surplus killing, with the hope that this sets the stage for our broader discussion on whether surplus killing can be considered fun.

The challenges of capturing wild prey

As Mech *et al.* (2015) explain, the central problem for carnivores when hunting is to kill without themselves being killed or injured. Hunting is an inherently risky business for just about any predator, and no less so for canids, which may take on prey animals much larger than themselves. There have been many cases where prey have killed wolves in self-defence (Mech *et al.* 2015). This is because prey species are typically equipped with well-developed defensive equipment and behaviour. Indeed, this is an example of the interplay that gives rise to an 'arms race' between prey trying to avoid being eaten and the wolves trying to eat them. This results in adaptations by the prey (e.g. physiological changes such as greater speed, better camouflage or higher rates of reproduction, but also behavioural changes such as increased vigilance, or frequenting areas of more structural complexity/cover – sometimes nicknamed 'the landscape of fear'– see Laundré *et al.* 2010), and in turn, changes in hunting strategies employed by wolves (Mech *et al.* 2015).

Most methods that predators use to capture prey are often not overly efficient, with capture rates usually varying from 4.5% to 10.8% (Robinson and Bolen 1989). Wolf hunting success is highly variable (as low as 1%, and as high as 56%), with rates varying according to the number of prey and predators involved (Mech and Peterson 2003). In dingoes, Thomson (1992) found that of 272 observed chases of kangaroos, only 25 (9.2%) were successful. Predators are faced with two problems; first, they must locate prey animals, and second, they must confront them. It may take several hours of travelling to locate prey, and more to locate prey they can actually kill. This of course depends greatly on the environment and species involved. Often, prey flee, so in the case of wolves, they must catch up to their fleeing prey, slow it down, inflict injuries, usually through thick hair and hide, and eventually disable the struggling prey, all whilst avoiding hooves and horns. In the case of some larger prey (such as bison), which often stand their ground, this forces the wolves to change tack by confronting them in-place and attempting to bite into them until they are incapacitated. These events make the wolves highly vulnerable to injury or death, forcing the ultimate choice of whether to risk trying to kill the prey (Mech *et al.* 2015).

As all the larger prey (with the exception, to an extent, of domestic varieties), possess alert senses and dangerous defences, they are neither easily found nor easily caught and killed. Each have specific defences and anti-predator behaviours that can be lethal to canids. It is for this reason that Mech *et al.* (2015) argue it would be "dangerous and foolhardy for wolves to 'kill for sport'" (p. 7). So the majority of the time, we can probably assume that catching and killing prey is arduous and dangerous work, and this probably makes surplus killing events relatively infrequent, at least when it comes to wild prey. There are some useful approaches to modelling predator–prey systems, and the selective forces behind hunting strategies, that will help us further cast surplus killing in an evolutionary light, which we will now briefly explore.

Optimal foraging theory and partial prey consumption

It is a reasonable assumption that, given all the challenges wild animals face to stay alive and successfully reproduce, being efficient at finding and securing food is of critical importance. We can imagine that the closer an animal is to 'optimal' at these and any associated tasks, the better off it will be. There are many assumptions attached to the models used to explore optimal foraging (conceptual models used to make predictions about how animals find food) but an important one in this case is that animals are assumed to maximise their energy intake (Dugatkin 2004). Further, how 'profitable' a prey item is can be taken to be the ratio of the amount of energy it provides/the amount of handling time (time it takes the predator to handle the food, beginning from the time the predator finds the prey item to the time the prey item is eaten; Dugatkin 2004). Thus, a more profitable prey item will have a relatively high amount of energy and require a relatively low amount of handling time. Alternatively, a 'patch' view sees animals making decisions about how long to stay in a particular area or use a particular resource. One of the most well cited patch-related models is the Marginal Value Theorem (Charnov 1976). The theorem predicts that an animal should leave a patch when energy intake drops below an average for that habitat.

As an example, compare dingoes killing sheep, with the emphasis here on killing (part of the amount of handling time in the above ratio) and not necessarily eating sheep, in comparison to kangaroos, which can be regarded as a preferred prey species (Thomson 1984a). Capturing and killing kangaroos in many circumstances would be expected to be more challenging for dingoes than capturing and killing sheep, not least because kangaroos are faster and more agile than sheep. They are also probably better at defending themselves. In stark contrast, it is possible that the handling part of the ratio involving sheep is so low that it bears almost no cost to dingoes (Thomson 1992). This makes the handling element of the ratio much higher for kangaroos than for sheep, yet it would appear that dingoes still prefer to eat kangaroos. Perhaps the energetic benefits of much of the sheep carcass is substantially lower than is the case for kangaroos? This is not clear; however, it is often noted that at least some portions of sheep are eaten, albeit often minimally. Are only the most energetically profitable morsels of sheep eaten? This is a rationale suggested by Zimmermann *et al.* (2015), who argue that an optimal strategy is to consume only the most nutritious parts of easily accessible prey while, for example, avoiding the risk of being detected by competitors or threats (e.g. humans). Of course, this is an oversimplification of optimal foraging theory. However, it serves as a useful lens with which to view the 'choice' dingoes have between sheep and kangaroos (or any other species) as potential prey items.

As is often the case, it is with wolves, rather than dingoes, that a most thorough examination of optimal foraging and surplus killing has been undertaken. As mentioned in our opening, Vucetich *et al.* (2012) are attempting to change the very conversation around surplus killing, starting with nomenclature. Rather than the arguably loaded term of surplus killing, Vucetich *et al.* (2012) advocate for a reframing of the behaviour in the context of what they see as a widespread practice exhibited by many species: that of partial prey consumption (PPC). After all, the term surplus killing has a tendency to imply wastage, gluttony and wantonness on the part of the carnivores involved (Vucetich *et al.* 2012). Further, Vucetich *et al.* (2012) contend that PPC may in fact represent an optimal foraging strategy, arguing that key predictions from previous theoretical assessments are that the average amount of carcass utilisation is negatively correlated with kill rate (i.e. as one increases the other tends to decrease) and utilisation is uncorrelated with the time between kills. This is exactly what they observed in wolves. Recalling the 'patch' optimal foraging model mentioned above, Vucetich *et al.* (2012) saw a carcass like a patch, and a predator must make a 'decision' (note, this does not necessarily imply a conscious decision) about how long to stay and eat the carcass or move on to find other food. They surmised that if catching other prey is relatively easy, and the energetic benefits of remaining at a carcass are low enough, then it is optimal to only consume a portion of the carcass and then move on. A very recognisable example of this involves bears consuming salmon. When salmon are in high abundance and relatively easy to catch, then each individual fish is only partially consumed, with a focus on the most energetic elements, like eggs in female fish and brains in male fish (Gende *et al.* 2001). The strategy revolves around maximising caloric intake. Conversely, as salmon abundance decreases, a greater proportion of the fish is consumed (Gende *et al.* 2001).

Such a strategy has obvious, potential survival benefits as bears are able to gain maximum weight in the shortest period of time, in preparation for periods without food (e.g. hibernation). Many other species, as well as more marginalised bears, are likely to benefit from scavenging fish remains so nothing is truly 'wasted'. In regard to the wolves that Vucetich *et al.* (2012) examined, moose (the principal prey) carcasses were largely consumed (median was 91%). Any occasions where significantly less consumption per carcass took place were relatively rare, but a proportionally large number occurred in a single year (1996) in which a particularly harsh winter was observed, and moose density was at its highest level in the 50 years over which observations had been taken. Abundance of moose declined by 80% during that same winter (Vucetich *et al.* 2012). The principle findings were that as kill rate increased, consumption per carcass decreased, and, that there was little observable relationship between consumption rate and the number of days between kills, as expected if PPC were an adaptive, optimal strategy (Vucetich *et al.* 2012).

In most instances, the probability of successful hunts is quite low. Thus, it is reasonable to assume that given the rare opportunity to kill or incapacitate large numbers of prey in a short space of time, there may be advantages to doing so for immediate gain, and also for long-term gain (i.e. utilising the resource in the weeks/months following the kill). There certainly does not seem to be an adaptive reason against surplus killing (Miller *et al.* 1985). Precariousness in hunting success might therefore give rise to an adaptive strategy to 'make hay while the sun is shining'. Even if kills spoil, nutrients may be gleaned from decaying carcasses for some time afterwards, and access to these may make all the difference in lean times. In cold climates, carcasses will also keep for considerably longer periods of time. In such circumstances, it might actually be disadvantageous not to kill as much as practicable, because at some point down the line, such resources might be vital to the survival of the individual involved, its offspring, or members of its social group, as Kruuk (1972) pointed out.

So there are some compelling evolutionary arguments in favour of predators undertaking surplus killing of wild prey, in at least some circumstances, but if the frequency of such events was high, it may also have a dramatic toll on prey populations. A major point from Vucetich *et al.* (2012) is that prey vulnerability may have changed in the severe winter observed in 1996, making moose easier to catch and requiring less energy expenditure per kill for wolves. This issue of prey vulnerability is worth examining more closely.

The vulnerability of free-ranging prey

As discussed earlier, the defensive strategies and weapons available to potential prey animals in thwarting predation events may be a major determinant of whether the outcome will end in death or injury for the prey animal (or the predator for that matter). Under normal circumstances, these traits work admirably, as the low kill success rates cited suggest. It is therefore relatively uncommon that predators get a chance to kill more prey than they can immediately consume, and further, for surplus killing to occur, such anti-predator defences must fail for some reason (Kruuk 1972). But what if such anti-predator strategies are not present?

For example, in relation to foxes involved in surplus killing events in Australia, Short *et al.* (2002) posited that the naivety of native prey to the threat of foxes made them much more susceptible to predation. This, they argued, was because there was no shared, evolutionary history between foxes as the predator, and native prey species. Short *et al.* (2002) outlined a number of case studies showing that particular native prey species, such as bettongs, bandicoots and wallabies were susceptible to surplus killing by foxes. Complicating this idea though has been research from Blumstein *et al.* (2009) demonstrating evidence for what they called the multi-predator hypothesis. This posits that provided there is at least some underlying risk of predation from any predators in a multi predator system, then prey species should retain some anti-predator responses or avoidance measures, even to predators that are not, or have never been, present. Independent support for this argument comes from Carthey and

Banks (2012), who showed that bandicoots tend to avoid the back yards of people with dogs, possibly because of stimuli such as dog urine associated with a potential predator with which they have some shared evolutionary history: the dingo. This avoidance was not observed in bandicoots on islands where dingoes were not present (Frank *et al.* 2016; see later in regard to Fraser Island, where both dingoes and bandicoots are present). Thus, in relation to the naivety concept raised by Short *et al.* (2002), it might have more to do with either that some anti-predator defences of prey (e.g. avoidance) do not transfer from dingoes to foxes, or that foxes, or environmental conditions, somehow circumvent or nullify these defences.

In one of the few published accounts involving dingoes killing native prey in apparent excess, Shepherd (1981) reported that in a seven-week period, five dingoes killed 83 mostly juvenile kangaroos near a dam. Dingoes only ate portions from about half the kangaroos killed. Large male kangaroos rarely fled in the presence of dingoes, whilst smaller kangaroos were usually highly vigilant and readily prone to flight responses near the dam. However, perhaps most intriguing of all, Shepherd reported that there were occasions when kangaroos of all sizes would pass by dingoes in the open, sometimes as close as five metres away, with neither the dingo nor the kangaroo showing much interest or concern.

From the single chase Shepherd (1981) observed, it appeared that although dingoes were capable of catching up with a fleeing kangaroo, a kangaroo could also outmanoeuvre dingoes to escape, as happened on the observed occasion. The vigilance that kangaroos also usually displayed suggested a recognition of dingoes as a threat, and interestingly, possibly even recognition of when dingoes were less of a threat. Autopsies did not reveal any disease or other debilitation that might have hampered kangaroo escape (Shepherd 1981). It is reasonable to assume therefore that something relating specifically to juveniles and their vulnerability, perhaps in combination with some unobserved tactic on the part of dingoes, temporarily tipped the scales in favour of dingoes.

Cases involving dingoes surplus killing other native species appear rare, perhaps because of there being at least some shared evolutionary history with common prey animals, in comparison to, say, cats and foxes. Consequently, most events where dingoes are implicated involve domestic animals like sheep (Short *et al.* 2002). Examining the dynamic between dingoes and native prey species in places absent of sheep therefore offers an opportunity to explore any resultant 'arms race' that might have developed. Fraser Island, which is home to some 100–200 resident dingoes (Jones and Appleby 2011) is the perfect natural laboratory in this sense, where it has been found that bandicoots are the most common species represented in dingo scats, although they appear to generally have a diverse diet (Behrendorff *et al.* 2016). So what is the association between dingoes and bandicoots? Angel (2006) conducted a preliminary examination of small mammal abundance in some parts of Fraser Island, and found that in certain areas, capture rates and associated abundance were quite high. Like Behrendorff *et al.* (2016), Angel (2006) identified bandicoots as a major source of dingo prey, with an assumed functional relationship approaching what is known as Type II,

This relationship is typified by a levelling off of predation when density of prey is very high, and was described by Dale *et al.* (1994) as the functional response of wolves to caribou in their study. A result of a possible Type II functional response is that as density of bandicoots increases, it becomes much easier for dingoes to catch them, but that at low densities, dingoes are still good at catching bandicoots, suggestive of a type of specialisation in hunting bandicoots for the island's dingoes. It is also possible that bandicoots have a reproductive strategy that somewhat accommodates relatively high rates of predation on Fraser Island, but this requires further investigation.

The temporary tipping of scales in favour of predators has been suggested in some published cases of surplus killing in wolves. For example, in North America, after a hard winter with deep snow, wolves might have an easier time killing prey that have been weakened, and in these settings, wolves may kill more prey than they can eat at that time. Here kill rates/success rates rise, and some surplus killing can occur (DelGiudice 1998; Mech *et al.* 2001; Vucetich *et al.* 2012). A harsh winter (1995–1996) was implicated in surplus killing events involving wolves in Minnesota and white-tailed deer (DelGiudice 1998). No similar events were observed in five previous winters (DelGiudice 1998). Examination of marrow fat in the femurs of killed deer led DelGiudice (1998) to conclude that poor nutrition and extremely poor condition of deer due to deep snow prevented them from getting to food, and these factors were the major determinants of surplus killing. Other studies lend support to the notion that severe winter conditions contribute to debilitated prey (e.g. Mech *et al.* 1971; Montgomery *et al.* 2013). But these severe conditions come few and far between, and they only occur during part of the year. Thus in essence, "they [wolves] kill as much as they can when killing is easy because such opportunities are rare, and if undisturbed, such surplus is entirely consumed" (Mech *et al.* 2015, 67).

There are also cases where higher rates of opportunistic killing by wolves have occurred during times when prey were highly vulnerable, but for reasons other than harsh winter conditions. For instance, Miller *et al.* (1985) confirmed that 34 newborn caribou calves had been killed by wolves within an area 3 km^2, apparently all within the same 24-hour period. Wolves had only partially consumed half of the carcasses. In what is a relatively rare occurrence, the authors actually observed wolves killing multiple caribou calves in quick succession. In one instance, a wolf was observed chasing down a caribou, caching it nearby (the wolf nosed snow over the carcass to partially bury it); it returned shortly after to chase caribou again but disappeared out of sight, and returned again to chase and kill another calf, briefly picking it up, before moving off to rest. Then for a final (observed) time, the wolf got up and again chased and killed a calf, before carrying it out of sight. All this occurred within a 28-minute period. On a separate occasion, a wolf was observed to kill three calves within a six-minute period (Miller *et al.* 1985). An interesting finding by Miller *et al.* (1985) was that of the calves partially consumed, all had no stomachs, and no milk curds were found near the scenes, lending weight to the possibility of highly specialised food selection by wolves, perhaps for energetic reasons.

These reports of surplus killing involving wild prey appear uncommon, and the factors that usually contribute to the vulnerability of wild prey are usually infrequent. Healthy, even partially grown, wild prey species tend to represent a formidable challenge to predators. But what about prey species that are almost always vulnerable to predation: domestic animals?

The vulnerability of domestic prey

In essence, domestic livestock represent the opposite of free-ranging prey in that, in principle, they are almost always highly vulnerable to attack from predators. Many breeds of livestock (but not all) have largely lost sufficient defensive aggression and related behaviours, or defensive weaponry, needed to thwart attack, and they remain only marginally capable of defending themselves (Lehner 1976). Almost all domestic animals have been bred for docility, but have retained the response of 'turning and running' from predators, a behaviour that mimics wild prey and stimulates attack from canids (Connolly *et al.* 1976). Allen and Fleming (2004) describe some of the behaviour of livestock that helps demonstrate a number of these vulnerabilities in the Australian dingo–livestock context. They note that sheep and goats will tend to flee when confronted by a dingo, sparking predatory chase behaviour. Sheep are easily outrun by dingoes, with one study observing high capture success rates (66%) of all chases initiated (Thomson 1992). Many observed chases also ended without capture as dingoes broke off to chase other sheep (Thomson 1992). Certain breeds of livestock may exhibit more effective anti-predator behaviour than others, and that can make some difference. Merinos (a breed of sheep) for instance, exhibit mobbing behaviour that leaves those sheep on the outer circle extremely susceptible to attack, whilst the inner sheep are less at risk (Allen and Fleming 2004). Further, livestock are vulnerable because they are usually confined to a limited space and cannot escape, unlike native prey that live in social groups where if an individual is taken, the rest can flee to safety. Lehner (1976) argues that surplus killing of livestock in coyotes is "not the result of a 'blood-thirsty killer', but an efficient predator which finds itself amidst an abundance of easy prey" (121). And, as Alderton (1994) adds, it is a natural response to the mass panic of the prey group (see also Biben 1979), disturbed by the predator that is unable to escape their confinement.

When describing hunting behaviour in the coyote, Lehner (1976) points out that chasing and killing are not causally linked to eating, and killing is not necessarily motivated by hunger. His observations support Kruuk's (1972) hypothesis that satiation (feeling full or satisfied) in carnivores "does not inhibit further catching and killing, but it probably does inhibit searching and hunting" (240). Yet when canids, such as the coyote, find themselves among easy prey (i.e. livestock), searching and hunting become superfluous (Lehner 1976).

At the beginning of this chapter, we highlighted some of the experiences of pastoralists in Australia in relation to surplus killing of sheep by dingoes. While there is relatively widespread anecdotal evidence of such events, there has been little empirical assessment. A primary, scientifically published source for dingoes surplus killing sheep comes from a study by Thomson (1992). Whilst it remains one of the most comprehensive studies on dingo behavioural ecology to-date, it was concentrated in the Fortescue River region of north-western Australia, so it is unclear how representative it is of other locations. Whilst Thomson (1992) concluded that some dingoes at this site engaged in surplus and 'excessive' killing, of both sheep and kangaroos, he did note that in relation to kangaroos killed, dingoes often returned to feed from carcasses they had previously been seen at. Cattle carcasses (it was not clear whether these were killed by dingoes, but many probably were not) on the other hand were rarely revisited and no data were presented for revisitation to sheep carcasses. The high number of successful chases of sheep Thomson (1992) observed, at approximately 66%, contrasts sharply with that for kangaroos (9.2%). One potentially important element of Thomson's (1992) observations was that sheep were actually attacked (rather than merely chased) more often when more than one dingo was involved. Thomson (1992) observed a total of 61 attacks on sheep by dingoes (solitary and in groups) in which eight sheep were killed outright, and a further 26 sheep appeared to sustain injuries. No timeframe was given over which these events occurred. Attacks were broken off by dingoes, not because sheep escaped or successfully defended themselves. Highlighting the vulnerability of sheep, Thomson (1992) observed dingoes feeding on sheep without having inflicted a killing bite.

With the exception of Thomson's (1992) observations, it is, therefore, generally unclear how widespread and under what circumstances surplus killing of sheep, or cattle for that matter, by dingoes actually takes place. On face value, given that livestock like sheep might often represent the largest, easily obtainable and relatively abundant prey available to wild canids in some areas, and given how vulnerable sheep are to attack, it is perhaps surprising that surplus killing of sheep does not occur at a very high frequency. It may be because canids actually prefer to hunt native prey over domestic prey, as dingoes certainly appear to prefer kangaroos over livestock (Allen 2014; Thomson 1984a; 1992). Indeed, prominent dietary studies of the dingo across Australia show that livestock (namely sheep and cattle) feature relatively uncommonly, even when readily available (Newsome *et al.* 1983; Corbett and Newsome 1987; Brook and Kutt 2011). Thomson's (1992) study was an exception in relation to sheep at one of his sites. It is also worth noting that analysis of scats and stomach content, the most common method for assessing diet, is not always a reliable means of actually assessing predation risk (Thomson 1992; Allen *et al.* 2012). An important point from Thomson (1984b) is that, at least in his study, kangaroo density was much higher in 'sheep country' than in unstocked areas. Thus, the higher availability of kangaroos in sheep grazing areas might be the very reason dingoes are attracted there in the first place.

There is an argument that canids "are not born predators of sheep; that is, sheep are not inherently recognized as a food source although they will elicit attack" (Lehner 1976, 121). The ability of predators to identify, capture, kill and consume prey is shaped through experience (Fox 1969). Corbett (2001) wondered whether the notion of search image, the idea that particular prey species register more readily than other potential prey species in the mind of the predator, played a role in dingo attacks on sheep. He pointed out that at Thomson's (1992) field sites, dingoes moving in from other areas took a long time to actually attack sheep. He described such dingoes as "inept hunters" of sheep, probably in the sense that they were inefficient at killing them, and often continued to kill and eat kangaroos, their preferred prey (Corbett 2001, 107). In contrast, dingoes observed by Corbett (2001) that were raised near sheep killed them almost exclusively, and rarely killed kangaroos. To summarise then, there are a raft of influences on the likelihood of canids attacking livestock, including the hunting experience of the canid, the population dynamics of the canid (social organisation and population), and the use of predator lethal control and other management strategies in the area (Lehner 1976). Keeping predators and domestic livestock apart and preventing or modifying what predators learn about domestic prey vulnerability might represent the clearest path to resolving conflict.

Behavioural, cognitive and neurological explanations

As we have discussed already, evolution plays an overarching, or ultimate role in causing behaviour, but this does not explain the more immediate, or proximate reasons why a behaviour is, or is not exhibited by an animal. All of the 'machinery' that has been selected for, including brains and nervous systems, sensory and endocrine systems, work together to control an animal's behaviour in a dynamic process. It is these systems to which we now turn our discussion.

A note on anthropomorphism and anthropodenial

Essentially, and perhaps not without a little irony, there are two schools of thought regarding the degree to which wild canids such as wolves and dingoes 'think' during hunting. There are those that warn against any hint of anthropomorphism, or the idea that animals are mistakenly ascribed with 'high-order' traits commonly attributed to humans; and there are those that argue that certain behaviours exhibited by animals clearly resemble high-order behaviour, similar to those also observed in humans. In relation to the latter, renowned animal behaviourist Frans de Waal (1999) viewed the opposite of anthropomorphism as anthropodenial (the refusal to attribute such high-order behaviours to animals, or indeed to see the animal in ourselves). For simplicity, we can view this as resulting in polarisation with a 'Cartesian' (named after Rene Descartes) or instinctual viewpoint at one end through to the 'cognitive' or intentional viewpoint at the other end, and of course, a spectrum in between. This is not the forum to delve too deeply into this apparent schism, but below, we attempt to provide some salient examples and points from both schools of thought, intermixed with some of our own opinions on the matter, specifically in relation to hunting behaviour and surplus killing. Readers interested in learning more of each perspective are also directed to Burghart (2004), Wynne (2007) and Horowitz and Bekoff (2007).

Fixed action patterns in relation to hunting

To many biologists and (particularly classical) ethologists, the behaviour of non-human animals is driven largely by intrinsic motor patterns, or species-specific, stereotyped products of natural selection (Lorenz 1982). Fixed action patterns are reflections of typical or average behaviour that involve complex chains of events that may play out over a considerable period of time. Biologist and canine expert Ray Coppinger argues that motor patterns are fundamental to understanding most, if not all of the behaviour of canids, including hunting behaviour. In relation to hunting, he posits five basic steps in what he calls the predatory sequence. Each of the steps involves a fixed-action pattern that leads to the fifth and final step (Coppinger and Coppinger 2002; Coppinger and Feinstein 2015; Lord *et al.* 2017). These steps are summarised in Table 11.1, and elegantly described by Coppinger and Feinstein (2015, 68):

> A wolf, having detected and then oriented toward a prey animal, first goes into the eye motor pattern, it stands or lies stock-still, staring fixedly at the prey. Vision here, is what is presumed to be the key input here that provides the releaser that triggers the motor pattern. The predator then moves into stalk, with its body lowered and its head down, directed toward to prey. Holding the prey in sight, the animal moves slowly forward. The next step in the chain is the chase. Its shifts into a full speed forward running movement. Two kinds of bite behaviour may then ensue. A grabbing bite that effectively disables the prey, and a killing bite that dispatches it. The carnivore then uses its carnassial and other teeth to dissect its victim. If all goes well, the sequence ends when predators consume their prey. This may also change slightly depending on the prey. For example, fore-foot stab might be released by small prey like mice rather than large prey [see Figure 11.1]. When this rule is engaged, wolves may substitute the fore-foot stab for chase, as well as head shake for the kill bite. This alternative pattern is clearly adaptive in the context of different types of prey.

Coppinger and Feinstein (2015) argue that the predatory sequence is intrinsic, stereotyped and automatic (almost like a behavioural 'rule book'), and thus no practice is required, and little learning is involved. They also question whether these predators have "any sort of intelligent grasp of their goals, or indeed, whether they are conscious of their own actions" (82). They suggest for instance, that wolves need to go through the entire predatory sequence, each aspect motivating the next, but that if the sequence is interrupted, the wolf cannot continue carrying out the kill (Coppinger and Feinstein 2015). The wolf and the dog (considered their domestic derivative), exhibit the same ancestral predatory motor patterns, but they display them at different frequencies and in different sequences (Lord *et al.* 2017). One way to show that such generalised motor patterns are inherited is by selecting for or against their expression in subsequent generations. To highlight this point, Coppinger

Table 11.1 Foraging behaviour motor patterns in wild canids, and breed-typical motor patterns in domestic dogs

Breed type	*Motor-pattern sequence*					
Wild canid	ORIENT>>	EYE>>	STALK>>	CHASE>>	GRAB-BITE>>	KILL BITE
Livestock guarding dog	(orient)	(eye)	(stalk)	(chase)	(grab-bite)	(kill-bite)
Herding dog (header)	**ORIENT>>**	**EYE>>**	**STALK>>**	**CHASE**	(grab-bite)	(kill-bite)
Healing dog (heeler)	**ORIENT>>**	eye	stalk	**CHASE>>**	**GRAB-BITE**	(kill-bite)
Retriever dog	**ORIENT>>**	eye	stalk	chase	**GRAB-BITE**	(kill-bite)

Source: Coppinger & Feinstein (2015, 90).

Notes: Behaviours in capitals occur in high frequency; lowercase indicates relatively infrequent or absent. Behaviours in parentheses are considered faults in show and competition dogs for a particular working task. The symbol >> indicates that the behaviours occur in a particular order and are always linked together in a particular order (associated patterns indicated in bold).

Figure 11.1 A coyote performing a forefoot stab, a typical predatory behaviour used by many species of canids when hunting small prey.

Source: Creative Commons/www.pixabay.com

and Feinstein (2015) used the border collie (a livestock herding dog) and in contrast, the Maremma (a livestock guarding dog). Border collies were intentionally bred to not express some predatory motor patterns (i.e. grab-bite is omitted, as is kill-bite, dissect and consume), but to express others (i.e. they eye, followed by stalk and then chase). The Maremma dog breed on the other hand, has none of the apparent predatory sequence, and if individuals do, they are considered 'faulty'. Coppinger and Feinstein (2015) further argue that it would be difficult to train border collies to eye sheep, setters to become point birds, and retrievers to retrieve (i.e. for these breeds to learn these behaviours). Rather, they are born knowing such behaviours. So much so that according to Coppinger and Feinstein (2015), cross-fostering experiments of border collies with Maremmas did not alter their breed-specific motor patterns.

The approach of researchers such as Coppinger and Feinstein (2015) moves the focus from the apparent goal or consequence of behaviour (in this case, the kill), to the process of the behaviour being carried out, which they see as innate. That is, the act of carrying out the motor patterns themselves is 'self-motivating' and 'self-rewarding' (see also section on neurological reward mechanisms). Coppinger and Coppinger (2002) believe that for a wild canid, hunting (i.e. carrying out the predatory motor sequence) is pleasurable, such that they will continue to seek out opportunities to display the sequence. They propose that a predator does not have to 'know' that hunting leads to eating at all, and question whether predators hunt specifically to eat or simply seek out opportunities to hunt which incidentally lead them to food. In support of this contention, they argue that "no [border] collie trialer or sled dog driver ever gives a food reward for performance and that the dog already got its reward by performing the instinctual behaviour" (Coppinger and Coppinger 2002, 202).

However, whilst it is a useful illustration of the types of behaviour often exhibited during hunting, the notion of hunting by canids being the result of fixed action patterns is probably of limited value. This is because it assumes, as the phrase itself implies, that hunting sequences are largely immutable (fixed), and inflexible to change in the face of new experiences and resultant learning. It makes sense that some important aspects of hunting (and other behaviours) are inherited, particularly in very stable and relatively unchanging circumstances, or when it is of dire survival importance. For example, some animals are born without any parental care and there are therefore important survival advantages to having a 'blueprint' of what to do (and not to do) in order to immediately survive. But canid species such as wolves and dingoes have a relatively long juvenile period complete with extensive parental and alloparental care. Learning and acquiring skills is therefore possible, and allows for considerable flexibility in hunting patterns. This is one potential reason why canids such as wolves have (or at least had) a large global distribution, from the arctic tundra, to deserts and most habitats in between. In this sense, being born with a set of innate responses specific to hunting one type of prey or in one circumstance might not be advantageous, while behavioural plasticity and an ability to learn might be very beneficial.

A broader ethogram of predatory behaviour

While elements of hunting or predatory sequences might be innate as suggested in the previous section, hunting behaviour consists of a complex range of behaviours, and exhibition of these behaviours usually needs to be flexible. Traditionally, it has been difficult to define hunting. This is in part because there is great variation in hunting techniques used across carnivores, and hunting tends to combine many behaviours. Further, simple definitions tend not to take into account variation in prey type or antipredator defences at various stages during predatory events, and ultimately a 'hunt' can refer to any one of several behaviours, individually and collectively (Creel and Creel 1995). Nonetheless, hunting behaviour has typically been divided across three major behavioural 'states' – *search, pursue* and *capture* (Holling 1965). Recently, MacNulty *et al.* (2007) sought to provide more detail to this rather simplistic explanation by conducting a detailed behavioural analysis of over 2,000 hours of observed wolf hunting behaviour. They carefully looked for patterns of behaviours within these observations, and determined that predatory behaviour occurs across six foraging states. Using their ethogram of foraging states (provided in Table 11.2), they describe a typical hunting event for wolves involving the following behaviours: Travel (*search*) > Approach/Stalk (*approach*) > Watch/Encounter/Circle herd (*watch*) > Attack herd/Harass/Chase group (*attack-group*) > Single-out/Rush/Chase/Pursue individual/Cut-off individual (*attack-individual*) > Grab/Seize/Attack/Physical contact (*capture*). They show that hunting strategy remains flexible, varying according to the context of each hunting opportunity. This includes the response to such variations as the prey type, dynamics of the prey encountered, makeup of the hunting party, prey vulnerability and behavioural reaction of the prey.

Cognitive aspects in relation to hunting

Beyond motor patterns and what might be considered 'rule-driven behaviours' it is difficult to ignore that in many instances where canids have been observed hunting, a strong argument can be made for higher-order cognitive processing

Table 11.2 Proposed ethogram of the behaviour of wolves hunting ungulate prey

Foraging state	*Definition*
Search	Travelling without fixating on and moving toward prey
Approach	Fixating on and travelling toward prey
Watch	Fixating on prey while not travelling (e.g. standing, sitting, crouching)
Attack-group	Running after a fleeing group or lunging at a standing group while glancing about at different group members (e.g. scanning)
Attack-individual	Running after or lunging at a solitary individual or a single member of a group while ignoring all other group members
Capture	Biting and restraining prey

Source: MacNulty *et al.* (2007, 597).

being involved. In one extraordinary anecdote for example, which Gunn *et al.* (2006) describe as a "specialized hunting strategy" (7), human hunters observed wolves running a caribou off a cliff, in the same way that humans have been recorded to do with other prey animals. A total of nine caribou were found grouped under the cliff in similar circumstances, suggesting it was no accident. Most of the carcasses had been heavily fed upon. There are also some similar reports, such as Nichols (2015), who described wolves exhibiting an attempt at cooperatively ambushing geese.

Intriguing observations of potential hunting strategies have also been observed in wild dingoes. For instance, a recent observation on Fraser Island provides a possible example of dingoes using water as a 'tool', where a dingo was observed to capture an echidna, bring it down to the ocean, dig a shallow hole, deposit the echidna in the hole and then, over a period of about an hour, drown it before eating it (L. Behrendorff, personal communication, 27 April 2017). This is not the first time dingoes have been reported using tools (Smith, Appleby and Litchfield 2012). Newsome (2001) stated that dingoes that selectively targeted female wallabies carrying pouch young that were often ejected when pursued, was akin to dingoes 'farming' wallabies. He also relayed observational evidence of dingoes seemingly using fencing at a dam to facilitate the capture of red kangaroos. Purcell (2010) observed four dingoes chasing an adult eastern grey kangaroo into a river, where they continued to attack it, in what appeared to be an attempt to both injure and drown the animal. In this particular example, the kangaroo escaped, but Purcell referenced another observer that had noted numerous kangaroo carcasses along the shoreline of the river, and believed that the same pack of dingoes used the water line and associated vegetation to improve their chances of capturing kangaroos. Similar anecdotes of wallabies being 'herded' into the ocean and drowned have been reported on Fraser Island (R. Appleby, personal observation).

Describing wolves hunting muskoxen (*Ovibos moschatus*), Mech (2007) concluded that a group of wolves with considerable experience hunting together that had learned each other's usual hunting approaches under various conditions, could easily employ highly specialised strategies. Such specialised strategies may require considerable intelligence in order to be properly executed. Mental processes that have been described for free ranging wolves hunting include 'insight' (Packard 2003), as well as 'foresight', 'understanding' and 'planning' (Mech 2007). These cognitive processes have also been attributed to captive wolves (Fox 1971) as well as captive dingoes (Smith, Appleby and Litchfield 2012). Mech (2007) defined insight as "the perceiving of a solution" (148), purposiveness as "deliberate behavior with an objective" (148) and planning as "deciding to behave in a way that considers information relevant to a perceived outcome" (146). Mech went further and said that the two observations he reported showed understanding (which he defined as "comprehending complex relationships", 146), and foresight ("behaving appropriately for dealing with a future event", 146). An interesting element of Mech's (2007) report is that he had not observed anything similar in a previous pack (Mech 2007), suggesting the possibility of innovation occurring in relation to this particular pack's hunting strategies. None of the examples here appears entirely explainable by 'simple' types of learning such as associative or trial-and-error learning (the latter being where the solution to a problem is more or less stumbled upon after repeated exposure) let alone from instinct in isolation, but some caution in this regard is nonetheless required (Shettleworth 2010). It is a tantalising prospect, however, to consider the possibility that wolves, and closely related canids such as the dingo, have the capacity to consider actions in the context of the future.

It is important to realise that not every anecdote or element of the observations described here is necessarily the result of 'high-order' behaviours, and simpler explanations might be more appropriate in at least some cases. For instance, caching, or the storing of food for later consumption, has been observed in a wide variety of species, including insects, so such 'forward planning' might be an innate, or hard-wired behaviour in many cases, rather than the product of cognitive decision making (Shettleworth 2010). Some have argued that a lack of detailed hunting observations precludes conclusions relating to more complex cognitive abilities, and that even when cooperative and collaborative hunting does occur, advanced cognition is not required in order to achieve it (Bailey *et al.* 2013). In fact in one case, Muro *et al.* (2011) describe simulations in which autonomous agents (simulated wolves) require no social structure and no communication in order to effectively hunt 'together', other than being faster than prey. Instead of collaborative and coordinated behaviour, Muro *et al.* (2011) describe two simple rules that wolf agents can follow in order to successfully hunt: (1) move toward prey until a minimum, safe distance is reached; then (2) move away from other wolves that are also close to prey (i.e. spread out, which results in encircling). Parsimony is a foundational element of science, so salient arguments that suggest rather simple explanations are at the heart of aspects of group hunts are naturally compelling. Arguably though, there is no reason that these rules could not apply to virtually any animal agent, including human agents, a species that we would readily attribute higher cognitive function to, provided that as in this example, they can similarly outpace and surround prey animals.

Muro *et al.* (2011) clearly state that it was not their intention to portray wolves as lacking the propensity for complex communication and cognitive abilities. However, the rules of Muro *et al.* (2011), risk evoking the outdated concept of animals as automata (largely unthinking or very simple rule-based machines). The concept of automata does not fit comfortably with the rich behavioural repertoires and complex social interactions often observed in gregarious canids and other animals, nor with emerging recognition of, for example, some animals as emotional and potentially conscious beings (Allen and Bekoff 1999; Boly *et al.* 2013; Griffin 2013), and cases of behaviour (e.g. Horowitz 2009; Krachun *et al.* 2016) that challenge simplistic conclusions. Of course, this does not mean that relatively simple rules born from, for example, hard-wired behaviour or associative learning, do not apply to aspects of hunting and many other instances of animal behaviour, as is also often the case with human behaviour (Hutchinson and Gigerenzer 2005; Shettleworth 2010). Mech (2007) too expressed reservations about attributing complex mental states to wolves, but he also provided descriptions of behaviour that belie the idea of wolves being something akin to automatons. One example was at the beginning of a hunt; one wolf was observed approaching another and nudging it with its nose upon which the second wolf got up and they moved off together to where they could both observe muskoxen. Perhaps there is also a simple explanation for such behaviour, another rule that could be generated. The problem is that hard-coded rules, which with the benefit of hindsight, can be encoded into agent behaviour, do not allow for the variation and spontaneity often described in wolf, and other wild canid hunts. Rules only work when nothing breaks them.

Nonetheless, if the two simple rules proposed by Muro *et al.* (2011) do provide a reasonable basis for explaining aspects of successful hunting behaviour, caution is obviously required in drawing conclusions linked to more sophisticated aspects of cognition, as has also been highlighted by Bailey *et al.* (2013). In this sense, the simplest possible answer, for example, drawn from associative learning, becomes the null hypothesis (Haselgrove 2016). However, we should not be too quick to assume simple rules or associative relationships between stimuli explain all non-human animal behaviour. Ultimately, more data are required in order to better determine underlying mechanisms controlling behaviour, including those related to hunting.

Neurological reward mechanisms

Pleasurable experiences occur in, and are mediated by the brain, in both animals and humans (Berridge and Kringelbach 2008). Reward is actually a composite of various, complex psychological components relating to 'liking' (the hedonic or pleasurable part of a reward), 'wanting' (the motivation for seeking a reward) and 'learning' (associations and predictions about rewards) that each occur in partly separate brain regions shared by most mammals (Berridge and Kringelbach 2008; Salomone and Correa 2013). Salomone and Correa (2013, 4) describe 'appetitive motivation' as 'seeking' behaviour, or the tendency to orient toward palatable stimuli, or "increasing the proximity, probability or availability of stimuli" as well as consumption. Similarly, Balcombe (2009) speaks of 'appetitive' or 'consummative' behaviours. The difference is that the former is motivated by desire (e.g. for mates, for food) and the latter by pleasure (e.g. mating, eating). In some ways, hunting generally fits in better with the first descriptor: desire for food drives an animal to hunt, more than it does with the latter, pleasurable act, which is more readily associated with eating (Balcombe 2009). However, we also cannot rule out potentially pleasurable elements of catching and subduing prey.

Evolution has resulted in responses to certain stimuli associated with survival and reproduction being pleasurable (rewarding). Two of the most fundamental involve mating and food (Berridge and Kringelbach 2008). In both humans and animals, there is extensive evidence that 'objective liking' can occur subconsciously, although 'subjective liking' occurs consciously (Berridge and Kringelbach 2008). The difference is that there is a consistency across species in the nature of behavioural responses to certain stimuli (e.g. 'enjoying' sweetness of a food) regardless of questions of consciousness (Berridge and Kringelbach 2008). The same is true for 'disliking' responses (e.g. evoked by bitter food), which are objectively different from 'liking' responses (Berridge and Kringelbach 2008).

The role of dopamine, long thought to be a principal mediator of hedonic pleasure, is still somewhat unclear. For instance, rather than being directly related to hedonic reward, it may have more to do with reinforcement and learning mechanisms, predictions about reward (Berridge and Kringelbach 2008) and the 'stamping-in' of memories associated with motivation importance of certain stimuli (Wise 2004; Berridge and Kringelbach 2008). In this sense, it appears to relate more with the 'wanting' component of reward (Berridge and Kringelbach 2008).

A major pathway through which reward mechanisms have been studied in animals is using food. It is often used as the reinforcer (a stimulus that increases the occurrence of a behaviour – note that this is operationally different from the notion of reward) in instrumental learning experiments. As we have pointed out though, the acts of hunting, killing and eating are discernibly different, such that the hedonic pleasure assumed to be associated with eating need not be present, or occur at the same scale or with the same valence, in hunting or killing. Research into any pleasure associated with hunting and killing from a neurological standpoint could not be found. However, some researchers assume that aspects of hunting (e.g. chasing) are potentially pleasurable. Recall for instance that according to Coppinger and Feinstein (2015), the act of carrying out the predatory sequence is the rewarding process, not the endpoint (that is, the kill). In support of this, Arons and Shoemaker (1992) showed that variability in predatory behaviour (response to a mouse, ball, conspecific and three sheep) across three pure-bred dog breeds (border collies, Shar Planinetz, Siberian huskies) correlated with neurochemical variability in specific brain regions and neurochemical characteristics. They discovered that across four brain regions, these three different types of working dogs significantly differed in the amount of dopamine found in their neural tissue. It is largely accepted that dopamine is a neurotransmitter capable of mediating general arousal and motor activity. Not surprisingly, frenetic and hyperactive border collies, which are expected to chase after and influence the direction and behaviour of sheep, show up to four times the dopamine level of the livestock guarding dogs, who are generally lethargic. It is these neurological and hormonal characteristics that are thought to underlie motor patterns, and thus some aspects of behaviour (Coppinger and Coppinger 1996). Again, however, we are aware of no comparable neurological examinations involving wild canids in relation to hunting or predatory behaviours, so strong conclusions relating to the pleasure derived from such behaviours cannot be made.

Can killing be considered leisure?

A great deal of the discussion relating to whether killing is 'fun' probably comes down to definitions, which is probably not unique in the study of animal behaviour. Consider for example, one definition of leisure: "a period of recharging the batteries or relaxation, before returning to work" (Carr in press). As we broached earlier, in an animal context, we might think of work as the more serious business of surviving and raising young, which we can assume makes up the vast majority of wild animals' lives. Apart from rest and sleep, an obvious parallel with relaxation and leisure exhibited by animals is play. Play tends to defy a succinct definition, not least because it lacks succinct agreement in terms of evolutionary function (Dugatkin 2004). A widely cited definition of play comes from Bekoff and Byers (1981: 300): "Play is all the motor activity performed postnatally that appears to be purposeless, in which motor patterns from other contexts may often be used in modified forms and altered temporal sequencing". "Purposeless", qualified with the preceding words "appears to be", does not infer useless or worthless, more that any purpose is not always immediately apparent. This definition of play though is not without problems, including the fact that, for example, stereotypic behaviours exhibited by zoo animals (e.g. pacing up and back along fencing) also meets this definition, despite no one readily equating this behaviour with play (Bekoff and Allen 1998). Presumably the same can be said for equating such behaviour with leisure.

McFarland (2014) actually defines play as a leisure activity more directly, with leisure itself described as those activities that are exhibited when demands on an animal's time, such as finding food, are not severe. Note that this ranks behaviours based upon how critical they are to survival and reproduction, with leisure and play being of a relatively low rank in this regard. This though, seems to be more of a description of what play and leisure are not, rather than what they are. Janik (2015, R7) states that definitions of play require that such behaviour should "not contribute to immediate survival needs" and be "intrinsically rewarding, pleasurable, spontaneous, voluntary and intentional". Janik (2015, R7) adds that play should occur when an "animal is relaxed – that is, when it is not immediately threatened or in a competitive situation". The act of hunting and killing, unlike play, is usually essential for survival. This somewhat muddies the water in terms of hunting meeting the criteria for what constitutes leisure, despite any pleasurable benefits derived from the activity.

As a final note on defining play, we are particularly fond of a description of play from Dugatikin (2004) that both sums up the quandary of defining play, and yet also how readily we all recognise it: "It seems that those who study play have adopted a definition similar to the United States Supreme Court's definition of pornography: they can't say exactly what it is, but they know it when they see it" (512).

We recognise that leisure, at least in humans, embodies a "vast and varied set of activities incorporating an incalculable range of personal interests, market provisions, and public resources" (Kelly 2012, 15). Unfortunately, definitions of leisure are primarily human-centric, so applications to non-human animals are potentially problematic. According to Kelly (2012), leisure is defined as "activity chosen in relative freedom for its qualities of satisfaction" (3), and is done in a "relatively unconstrained and uncoerced manner. It is done freely" (9). Leisure encompasses activity that is "chosen more for its own sake than for ends related to survival or necessity" (9). Note the parallel of the latter, in-part, with the apparent purposelessness of play described by Bekoff and Byers (1981).

So can we use the concepts of leisure and play interchangeably, and even with the more subjective term of 'fun'? We might readily equate the terms, but we obviously need to be cautious about ascribing internal states like fun and enjoyment to animals if we have no way to determine these states. At the same time, we suspect that at least some aspects of play, and other behavioural states in animals can be equated with fun, and producing rewarding experiences, as we suggested in the section on neurological reward mechanisms. Certainly, some of the leading experts in relation to animal play recognise that it is likely to be fun for the animals engaging in it (Biben 1998; Spinka *et al.* 2001; Allen and Bekoff 2005). In fact, an entire issue of the prestigious science journal *Current Biology* (Volume 25 Issue 1, January 2015) was devoted to the biology of fun and included many contributions from animal behaviour researchers. Canid behavioural expert Marc Bekoff (2015) wrote an article about play and fun in dogs in the same issue. He defined fun as: "doing something amusing, enjoyable, and pleasurable and feeling good about it" (Bekoff 2015, R5). It seems reasonable then to assume that there is some overlap between play and fun.

Another definition of leisure from Carr (in press): that "leisure is about learning" appears to marry well with one function of play; that of learning and training for the future. Note how it is in the future that play is then seen to become more purposeful. The notion that "leisure is a social activity" (Carr in press) also fits well with the concept of social play. Play also often involves inhibited (e.g. biting) or exaggerated (e.g. pouncing) behaviours that resemble and indeed draw from more 'serious' behaviours such as fighting and hunting, albeit often out of order (Biben 1979; Bekoff and Byers 1981). While we do not want to suggest that there is a direct link between play and surplus killing, because of the similarity in exhibited behaviour there are probable links between play fighting as practice for real fighting, and play hunting as practice for real hunting, so studying play behaviours resembling hunting motor patterns might offer some useful insights as to how surplus killing might arise as animals develop. Of more immediate concern to us here, however, is whether surplus killing fits as well with the notion of leisure activity, integral with notions of sport (Carr in press) and practice, as play appears to.

Perhaps the most in-depth experimental assessment of the relationship between play behaviour, hunger, killing and surplus killing comes not from canids, but from another species that has found kinship with humans, the domestic cat. The propensity for cats to 'play' with their prey is a well-known anecdote, but it was not until Biben (1979) examined play and predatory behaviour in the cat that this was experimentally assessed in detail. Biben (1979) used an operational definition of play, describing it as any behaviour that involved a cat's attention being directed at prey, or where the prey is touched or manipulated in some way without injury. Biben (1979) described such play behaviour as inessential to predation because mice, the usual prey of cats, could have been killed with minimal effort during bouts of play. Although play behaviour was not directly part of killing sequences observed, and the two sets of behaviours were easily distinguished from one another, Biben (1979) found a positive association between them. Whilst surplus killing was observed, 84% of the carcasses of mice were consumed by cats entirely or partially. Large differences in individual cat behaviour, such as their relative experience in killing, appeared to be an influential factor. Biben (1979) concluded that killing sprees (surplus killing) were most likely a consequence of a predator becoming over-excited by panicked prey that were easily captured, because, for example, they were confined (this is akin to the behaviour of canids surrounded by sheep described earlier). However, an important interpretation that can be made on the basis of Biben's (1979) results is that the act of killing or injuring itself was, very rarely (and probably only accidentally), also an act of play.

Conclusion

We began this chapter with the question of whether surplus killing was fun, and by approximation, could it be considered a leisure activity. On the surface, it probably seemed like it would be a relatively straightforward question to answer. Hopefully, we have demonstrated that it is actually far from it.

For instance, we discussed that there are potentially neurological reward mechanisms, such as the production of dopamine associated with certain behaviours, or other similar processes, that reinforce certain behaviours such as feeding, and separately, possibly elements of hunting as well. If such reward mechanisms play a role in the exhibition of surplus killing, and in turn, we equate these reward mechanisms with 'pleasure', then we might conclude surplus killing is, at least in-part, 'fun'. A major point here though is that it is far from clear whether this is actually the case from an empirical standpoint. Also, by this measure, a vast array of behaviours also probably meet the same criteria.

Consider, for example, the idea of an animal mating for 'fun'. Both mating and killing have ultimate (evolutionary) reasons to explain their occurrence, but there are also a number of proximate ones, including potentially, pleasure. After all, we would hypothesise that it is not likely to be the production of offspring (ultimate causation) utmost in the minds of animals when they seek out mates. It is more likely that proximate reasons motivate the behaviour, including hormones that influence neurological processes, or the 'wanting' (motivation to seek a reward) component described by Berridge and Kringelbach (2008). So do animals mate for fun? In the proximate sense we could potentially say yes. Yet many wild canids only mate seasonally, so does that mean it is only fun mating in mating season? And if mating does not result in offspring, should we classify it as 'surplus' mating?

An analogous follow-up question we asked ourselves in relation to surplus killing was: if it is really that fun, would it not happen a lot more? Of course, there might be very good reasons why it does not happen more often that have nothing to do with canid motivation (e.g. management and livestock husbandry methods). However, in Thomson's (1992) study, there seemed fewer restrictions and management approaches to prevent dingoes attacking sheep (as this was part of the experiment), and sheep were also very easy to kill compared to kangaroos. We might have therefore expected that if fun is what motivates surplus killing, it should have happened very frequently. Thomson (1992) of course did report cases of surplus killing, and we certainly recognise that dingoes and other wild canids can exact a severe toll on some sheep farming enterprises. Our only point is to wonder why virtually any vulnerable sheep should remain alive or uninjured in an area for any real length of time if killing them was such great fun.

Arguably, there are also some parallels between hunting and killing and the notion of work (as opposed to leisure), which we might not be as quick to equate with 'fun'. The often difficult job of staying alive and successfully raising young, the relatively low success rates of hunting attempts for wild canids, the constant threat from competitors and/or potential predators, haphazard environmental conditions, the dangerous task of successfully chasing and killing formidable prey animals and the implications of evolutionary arms races show just how serious hunting is: it is literally life and death. In this sense, any pleasure derived during hunting is probably more akin to the notion of job satisfaction, rather than the fun had during downtime.

We suspect that when anyone suggests it is 'fun' or 'sport' for predators to kill or injure animals in excess of their immediate needs, they are potentially engaging in anthropomorphism, but depending on the way it is framed, this may not necessarily be a bad thing. For instance, it may be more in line with what Burghardt (2007) calls 'critical anthropomorphism'. Ascribing emotional, mentalistic elements to animals for which there is no definitive, supporting evidence, or reasonable way to attain supporting evidence, might be called uncritical anthropomorphism. Posed as a statement, the notion that 'killing is fun', without supporting evidence, drifts into unscientific territory. In fact, it may drift into more of a philosophical arena regarding something akin to the moral responsibility of animals. However, posed as a preparatory hypothesis, 'is killing fun?' it becomes a complex, challenging and valuable starting point to explore a variety of questions. As the preface for formalising hypotheses, it stimulates ideas that can then be tested empirically (Burghardt 2007), which is a foundation of scientific enquiry. Perhaps then, an interesting point about an unqualified statement that killing in excess of immediate needs is fun, is that it might actually say more about us as observers, and how we think, than it does about the observed.

In a similar vein, we touched on some of the ongoing debate regarding the role cognition plays in the behaviours exhibited by animals, including hunting and killing. The foundational aspect of parsimony in science has led some to preclude attribution of higher cognitive functioning to non-human animals, in favour of possible explanations relating to 'simpler' processes, such as associative learning. Such researchers rightly warn against the misattribution of cognitive prowess where evidence is scant. Others though, including ourselves, see the shared challenges and problems faced by all animals, including humans, the many similarities in brain and physiological structures that have resulted, and homologies in behavioural expression, as a reasonable basis for wondering about higher cognitive abilities in other animals. Even species with which humans do not share a recent ancestor (e.g. birds; cephalopods) appear to possess remarkable abilities in solving problems, suggesting the compelling possibility that high-order cognition might have independently arisen on multiple occasions, just as other remarkable characteristics like flight or sight have. Proponents of this view recognise that whilst it is important to avoid uncritical anthropomorphism, neither should we succumb to anthropodenial, lest progress in our understanding be potentially stifled. Arguably, the result of such a debate will have a great bearing on how we view animals and their behaviour generally, and indeed, whether it is possible to attribute intentional states to canids is likely to influence how a phenomenon such as surplus killing is viewed.

In sum, there are potentially elements of pleasure in the act of hunting and killing prey. But on the whole, hunting is a dangerous and risky endeavour, and one that is an essential behaviour required for survival. In this sense, using existing definitions of leisure (albeit human-centric ones), killing prey cannot be considered a leisure activity, as say, play might be. Killing in excess of need (surplus killing) adds an additional element to the issue. After satiation, are animals killing merely because the act is pleasurable, or is it due to other evolutionary forces and happenstance? Ultimately, we think more empirical data are required to definitively answer this. For now, we err on the side of caution, and suggest that the act of surplus killing probably does not represent predators such as wild canids simply having fun, but a logical result of an evolutionary arms race, where on rare occasions, the predator temporarily comes out the overwhelming victor.

References

Alderton, D. 1994. *Foxes, Wolves and Wild Dogs of the World.* London: Blanford Press.

Allen, C., and Bekoff, M. 1999. *Species of Mind: The Philosophy and Biology of Cognitive Ethology.* Cambridge, MA: MIT Press.

Allen, C., and Bekoff, M. 2005. Animal play and the evolution of morality: an ethological approach. *Topoi* 24 (2): 125–135.

Allen, L. 2014. Wild dog control impacts on calf wastage in extensive beef cattle enterprises. *Animal Production Science* 54: 214–220.

Allen, L., and Fleming, P. 2004. Review of canid management in Australia for the protection of livestock and wildlife-potential application to coyote management. *Sheep & Goat Research Journal* 19: 97–104.

Allen, L., Goullet, M., and Palmer, R. 2012. The diet of the dingo (Canis lupus dingo and hybrids) in north-eastern Australia: a supplement to the paper of Brook and Kutt (2011). *The Rangeland Journal* 34 (2): 211–217.

Angel-E, D. 2006. Dingo diet and prey availability on Fraser Island. Masters dissertation, University of the Sunshine Coast, Queensland, Australia.

Arons, C., and Shoemaker, W. 1992. The distribution of catecholamines and β-endorphin in the brains of three behaviorally distinct breeds of dogs and their F 1 hybrids. *Brain Research* 594 (1): 31–39.

Bailey, I., Myatt, J., and Wilson, A. 2013. Group hunting within the Carnivora: physiological, cognitive and environmental influences on strategy and cooperation. *Behavioral Ecology and Sociobiology* 67 (1): 1–17.

Balcombe, J. 2009. Animal pleasure and its moral significance. *Applied Animal Behaviour Science* 118: 208–216.

Behrendorff, L., Leung, L., McKinnon, A., Hanger, J., Belonje, G.,Tapply, J., Jones, D., and Allen, B. (2016). Insects for breakfast and whales for dinner: the diet and body condition of dingoes on Fraser Island (K'gari). *Scientific Reports* 6: 23469. doi:10.1038/SREP23469

Bekoff, M. 2015. Playful fun in dogs. *Current Biology* 25 (1): R4–R7.

Bekoff, M., and Allen, C. 1998. Intentional communication and social play: how and why animals negotiate and agree to play. In *Animal Play: Evolutionary, Comparative, and Ecological Perspectives*, edited by M. Bekoff and J. Byers, 97–114. Cambridge and New York: Cambridge University Press.

Bekoff, M., and Byers, J. 1981. A critical reanalysis of the ontogeny and phylogeny of mammalian play: an ethological hornet's nest. In *Behavioral Development*, edited by K. Immelman, G. Barlow, L. Petrinivich and M. Main, 296–337. Cambridge: Cambridge University Press.

Berridge, K., and Kringelbach, M. 2008. Affective neuroscience of pleasure: reward in humans and animals. *Psychopharmacology* 199 (3): 457–480.

Biben, M. 1979. Predation and predatory play behaviour of domestic cats. *Animal Behaviour* 27 (1): 81–94.

Biben, M. 1998. Squirrel monkey playfighting: making the case for a cognitive training function for play. In *Animal Play: Evolutionary, Comparative and Ecological Perspectives*, edited by M Bekoff and J Byers, 161–182. Cambridge: Cambridge University Press.

Blumstein, D., Ferando, E., and Stankowich, T. 2009. A test of the multipredator hypothesis: yellow-bellied marmots respond fearfully to the sight of novel and extinct predators. *Animal Behaviour* 78 (4): 873–878.

Boly, M., Seth, A. K., Wilke, M., Ingmundson, P., Baars, B., Laureys, S., Edelman, D., and Tsuchiya, N. 2013. Consciousness in humans and non-human animals: recent advances and future directions. *Frontiers in Psychology* 4: 625. doi.org/10.3389/fpsyg.2013.00625

Brook, L., and Kutt, A. 2011. The diet of the dingo (*Canis lupus dingo*) in north-eastern Australia with comments on its conservation implications. *The Rangeland Journal* 33 (1): 79–85.

Burghardt, G. 2007. Critical anthropomorphism, uncritical anthropocentrism, and naïve nominalism. *Comparative Cognition and Behavior Reviews* 2: 136–138.

Carr, N., in press. Meaning of leisure. In *Leisure from International Voices*, edited by K. Henderson and A. Sivan. Urbana, IL: Sagamore Publishing.

Carthey, A., and Banks, P. 2012. When does an alien become a native species? A vulnerable native mammal recognizes and responds to its long-term alien predator. *PLoS One* 7 (2): e31804. doi.org/10.1371/journal.pone.0031804

Cathles, H. 2001. A landholder perspective. In *A Symposium on the Dingo*, edited by C. Dickman and D. Lunney, 75–83. Mosman, NSW: Royal Zoological Society of New South Wales.

Charnov, E. 1976. Optimal foraging, the marginal value theorem. *Theoretical Population Biology* 9 (2): 129–136.

Connolly, G., Timm, R., Howard, W., and Longhurst, W. 1976. Sheep killing behavior of captive coyotes. *Journal of Wildlife Management* 40 (3): 400–407.

Coppinger, R., and Coppinger, L. 1996. Biologic bases of behavior of domestic dog breeds. In *Readings in Companion Animal Behavior*, edited by V. Voith and P. Borchelt, 9–18. Trenton NJ: Veterinary Learning Systems.

Coppinger, R., and Coppinger, L. 2002. *Dogs: A Startling New Understanding of Canine Origin, Behavior and Evolution*. Chicago: The University of Chicago Press.

Coppinger, R., and M. Feinstein. 2015. *How Dogs Work*. Chicago: University of Chicago Press.

Corbett, L. 2001. *The Dingo in Australia and Asia*. Marleston, South Australia: J. B. Books.

Corbett, L., and Newsome, A. 1987. The feeding ecology of the dingo. *Oecologia* 74 (2): 215–227.

Creel, S., and Creel, N. 1995. Communal hunting and pack size in African wild dogs, *Lycaon pictus*. *Animal Behaviour* 50: 1325–1339.

Dale, B., Adams, L., and Bowyer, R. 1994. Functional response of wolves preying on barren-ground caribou in a multiple-prey ecosystem. *Journal of Animal Ecology* 63 (3): 644–652.

Dawkins, R., and Krebs, J. 1979. Arms races between and within species. *Proceedings of the Royal Society of London B: Biological Sciences* 205 (1161): 489–511.

de Waal, F. 1999. Anthropomorphism and anthropodenial: consistency in our thinking about humans and other animals. *Philosophical Topics* 27 (1): 255–280.

DelGiudice, G. 1998. Surplus killing of white-tailed deer by wolves in northcentral Minnesota. *Journal of Mammalogy* 79 (1): 227–235.

Dugatkin, L. 2004. *Principles of Animal Behavior*. New York: W. W. Norton.

Fox, M. 1969. Ontogeny of prey-killing behavior in Canidae. *Behaviour* 35 (3/4): 259–272.

Fox, M. 1971. *Behaviour of Wolves, Dogs and Related Canids*. New York: Harper and Row Publishers.

Frank, A., Carthey, A., and Banks, P. 2016. Does historical coexistence with dingoes explain current avoidance of domestic dogs? Island bandicoots are naïve to dogs, unlike their mainland counterparts. *PloS One* 11 (9): e0161447. doi.org/10.1371/journal.pone.0161447

Gende, S., Quinn, T., and Willson, M. 2001. Consumption choice by bears feeding on salmon. *Oecologia* 127 (3): 372–382.

Griffin, D. 2013. *Animal Minds: Beyond Cognition to Consciousness*. Chicago: University of Chicago Press.

Gunn, A., Miller, F., Barry, S., and Buchan, A. 2006. A near-total decline in caribou on Prince of Wales, Somerset, and Russell islands, Canadian Arctic. *Arctic* 59 (1): 1–13.

Haselgrove, M. 2016. Overcoming associative learning. *Journal of Comparative Psychology* 130 (3): 226.

Holling, C. 1965. The functional response of predators to prey density and its role in mimicry and population regulation. *Memoirs of the Entomological Society of Canada* 45 (S48): 1–62.

Horowitz, A. 2009. Attention to attention in domestic dog (*Canis familiaris*) dyadic play. *Animal Cognition* 12 (1): 107–118.

Horowitz, A., and Bekoff, M. 2007. Naturalizing anthropomorphism: behavioral prompts to our humanizing of animals. *Anthrozoös* 20 (1): 23–35.

Hutchinson, J., and Gigerenzer, G. 2005. Simple heuristics and rules of thumb: where psychologists and behavioural biologists might meet. *Behavioural Processes* 69 (2): 97–124.

Janik, V. 2015. Play in dolphins. *Current Biology* 25 (1): R7–R8.

Jones, D., and Appleby, R. G. 2011. *Preliminary Analysis of a Dingo Capture-Mark-Recapture Experiment on Fraser Island Conducted by the Queensland Parks and Wildlife Service*. Unpublished report. Brisbane, Qld: Griffith University, July.

Kelly, J. 2012. *Leisure* (Fourth edition). Urbana, IL: Sagamore Publishing LLC.

Krachun, C., Lurz, R., Russell, J., and Hopkins, W. 2016. Smoke and mirrors: testing the scope of chimpanzees' appearance–reality understanding. *Cognition* 150: 53–67.

Kruuk, H. 1972. Surplus killing by carnivores. *Journal of Zoology* 166 (2): 233–244.

Laundré, J., Hernández, L., and Ripple, W. 2010. The landscape of fear: ecological implications of being afraid. *Open Ecology Journal* 3: 1–7.

Lehner, P. 1976. Coyote behaviour: implications for management. *Wildlife Society Bulletin* 4: 120–126.

Lord, K., Schneider, R., and Coppinger, R. 2017. Evolution of working dogs. In *The Domestic Dog: Its Evolution, Behavior and Interactions with People* (Second Edition), edited by J. Serpell, 42–66. Cambridge: Cambridge University Press.

Lorenz, K. 1982. *The Foundations of Ethology: The Principle Ideas and Discoveries in Animal Behavior*. New York: Simon and Schuster.

MacNulty, D., Mech, L., and Smith, D. 2007. A proposed ethogram of large-carnivore predatory behavior, exemplified by the wolf. *Journal of Mammalogy* 88 (3): 595–605.

McFarland, D. 2014. *A Dictionary of Animal Behaviour* (Second Edition). Oxford: Oxford University Press. DOI:10.1093/acref/9780191761577.001.0001

Mech, L. 2007. Possible use of foresight, understanding, and planning by wolves hunting muskoxen. *Arctic* 60 (2): 145–149.

Mech, L., and Peterson, R. 2003. Wolf-prey relations. In *Wolves: Behaviour, Ecology, and Conservation*, edited by L. D. Mech and L. Boitani, 131–157. Chicago: University of Chicago Press.

Mech, L., Frenzel, L., and Karns, P. 1971. *The Effect of Snow Conditions on the Vulnerability of White-Tailed Deer to Wolf Predation*. USDA Forest Service Research Paper NC 52, 51–59.

Mech, L., Smith, D., Murphy, K., and MacNulty, D. 2001. Winter severity and wolf predation on a formerly wolf-free elk herd. *Journal of Wildlife Management* 65 (4): 998–1003.

Mech, L., Smith, D., and MacNulty, D. 2015. *Wolves on the Hunt: The Behaviour of Wolves Hunting Wild Prey*. Chicago: The University of Chicago Press.

Miller, F., Gunn, A., and Broughton, E. 1985. Surplus killing as exemplified by wolf predation on newborn caribou. *Canadian Journal of Zoology* 63 (2): 295–300.

Montgomery, R., Vucetich, J., Peterson, R., Roloff, G., and Millenbah, K. 2013. The influence of winter severity, predation and senescence on moose habitat use. *Journal of Animal Ecology* 82 (2): 301–309.

Muro, C., Escobedo, R., Spector, L., and Coppinger, R. 2011. Wolf-pack (*Canis lupus*) hunting strategies emerge from simple rules in computational simulations. *Behavioural Processes* 88 (3): 192–197.

Newsome, A. 2001. The biology and ecology of the dingo. In *A Symposium on the Dingo*, edited by C. R. Dickman and D. Lunney, 20–33. Sydney, Australia: Royal Zoological Society of New South Wales.

Newsome, A., Corbett, L., Catling, P., and Burt, R. 1983. The feeding ecology of the dingo. 1. Stomach contents from trapping in south-eastern Australia, and the non-target wildlife also caught in dingo traps. *Wildlife Research* 10 (3): 477–486.

Nichols, T. C. (2015). Cooperative hunting of Canada geese (*Branta canadensis*) by gray wolves (*Canis lupus*) in northern Quebec. *The Canadian Field-Naturalist*, 129(3), 290–292.

Packard, J. 2003. Wolf behavior: reproductive, social and intelligent. In *Wolves: Behavior, Ecology, and Conservation*, edited by L. D. Mech and L. Boitani, 35–65. Chicago: University of Chicago Press.

Purcell, B. 2010. A novel observation of dingoes (*Canis lupus dingo*) attacking a swimming eastern grey kangaroo (*Macropus giganteus*). *Australian Mammalogy* 32 (2): 201–204.

Robinson, W., and Bolen, E. (Eds) 1989. *Wildlife Ecology and Management* (Second Edition). New York: Macmillan Publishing.

Salamone, J., and Correa, M. 2013. Dopamine and food addiction: lexicon badly needed. *Biological Psychiatry* 73 (9): e15–e24.

Shepherd, N. 1981. Predation of red kangaroos, *Macropus rufus, by the dingo, Canis familiaris dingo (Blumenbach) in North-Western New South Wales. Wildlife Research* 8 (2): 255–262.

Shettleworth, S. 2010. Clever animals and killjoy explanations in comparative psychology. *Trends in Cognitive Sciences* 14 (11): 477–481.

Short, J., Kinnear, J., and Robley, A. 2002. Surplus killing by introduced predators in Australia – evidence for ineffective anti-predator adaptations in native prey species? *Biological Conservation* 103 (3): 283–301.

Smith, B. 2015. *The Dingo Debate: Origins, Behaviour and Conservation.* Clayton Bay, Victoria: CSIRO Publishing.

Smith, B., Appleby, R., and Litchfield, C. 2012. Spontaneous tool-use: an observation of a dingo (*Canis dingo*) using a table to access an out-of-reach food reward. *Behavioural Processes* 89 (3): 219–224.

Spinka, M., Newberry, R., and Bekoff, M. 2001. Mammalian play: training for the unexpected. *The Quarterly Review of Biology* 76 (2): 141–168.

Thomson, P. 1984a. Dingoes and sheep in pastoral areas. *Journal of Agriculture Western Australia* 25, 27–31.

Thomson, P. 1984b. Use of buffer zones in dingo control. *Journal of Agriculture Western Australia* 25, 32–33.

Thomson, P. 1992. The behavioural ecology of dingoes in north-western Australia. III. Hunting and feeding behaviour, and diet. *Wildlife Research* 19 (5): 531–541.

Treves, A., and Karanth, K. 2003. Human-carnivore conflict and perspectives on carnivore management worldwide. *Conservation Biology* 17 (6): 1491–1499.

Vucetich, J., Vucetich, L., and Peterson, R. 2012. The causes and consequences of partial prey consumption by wolves preying on moose. *Behavioral Ecology and Sociobiology* 66 (2): 295–303.

Wise, R. 2004. Dopamine, learning and motivation. *Nature Reviews Neuroscience* 5 (6): 483–494.

Wynne, C. 2007. What are animals? Why anthropomorphism is still not a scientific approach to behavior. *Comparative Cognition & Behavior Reviews* 2: 125–135.

Zimmermann, B., Sand, H., Wabakken, P., Liberg, O., and Andreassen, H. P. 2015. Predator-dependent functional response in wolves: from food limitation to surplus killing. *Journal of Animal Ecology* 84 (1): 102–112.

12 Ferals or food?

Does hunting have a role in ethical food consumption in Australia?

Heather J. Bray, Sebastian Konyn, Yvette Wijnandts and Rachel A. Ankeny

Introduction

There is increasing awareness of 'ethical' food production and consumption in most of the more developed countries, including Australia, but extremely different ideas exist about what is considered to be 'ethical'. Although hunting as a leisure activity is pursued only by a small percentage of the Australian population, even within these groups there are radically different motivations for and understandings of this activity. In this chapter, we explore the motivations underlying hunting among Australians via a review of scholarly literature, media articles from within Australia, as well as industry, government and regulatory body reports, and show how at least some hunters see their activities as consistent with ethical consumption, for instance because it promotes sustainability, eating local, controlling non-indigenous and feral animals and/or taking greater responsibility for animal death. Our findings reveal the complexities associated with hunting practices and underscore the need to pursue a deeper understanding of the values and motivations associated with these practices to facilitate a constructive dialogue about the role of hunting in food production and consumption in Australia.

Hunting as a leisure activity

Even a brief exploration of the scholarly literature reveals divergent terminology about, and rationales related to, hunting. In the broadest sense, a distinction is made between recreational and subsistence hunting (i.e., hunting for food necessary for survival). Recreational hunting includes sport hunting, which is said to be motivated by "the joy and thrill of hunting itself!" (Vitali 1990, 73) and for which the development of hunting skills is of primary importance (Leader-Williams 2009; Wade 1990). It also includes nature hunting, which fosters an "intimate experience of the complexity of ecological relationships and dependencies within a natural context" (Kellert 1996, in Simpson and Cain 2000, 185).

Hunting as a leisure activity is seen as a temporary return to nature, as "a 'vacation' or a diversion from the hunter's predicament as a civilized being subject to the constraints of history, culture, and community. We hunt in order to distance ourselves from our humanity" (King 2010, 151). The use of the word 'predicament' is notable here, and presents a view of hunters as individuals constrained or misunderstood by a society with radically different values. Hunting can provide opportunities to "get away from the hustle and bustle of everyday living" (internet comment, in Adams 2013, 49) to have "a purpose to be in the bush" (Steve, a hunter, in Marx 2012). This shift to hunting as a form of wildlife recreation arose in the latter decades of the 20th century, as urban dwellers took up hunting to reconnect with nature and to escape the pressures of city life (Franklin 1996). Hunting encourages individuals to go into the wild, connecting them far more intimately than other forms of involvement with nature (Kover 2010; Simpson and Cain 2000). As John, a duck hunter, states: "It's tradition to put food on the table. There's nothing better than connecting with the environment . . .. It might be the 21st century, but we're still hunter-gatherers. It's part of the life cycle" (in Munro 2012). However, the practices of contemporary hunters are fundamentally different to those of hunter-gatherer societies, despite attempts to utilise hunter-gatherers as a rationale in hunting discourses (Reis 2009).

The relationship between hunting and killing is complex. It is almost impossible to read anything on hunting as a leisure activity without coming across the work of Ortega y Gasset, especially his assertion that "one does not hunt in order to kill; on the contrary, one kills in order to have hunted" (Ortega y Gasset 1995, in Reis 2009, 584). However, the idea of killing as a necessary component of the hunt is contested by the work of Reis with hunters in New Zealand, who found that hunting for them was much more about the overall experience, and therefore they "need not kill in order to have hunted. The kill is just an episode within the hunting experience" (Reis 2009, 584). This attitude is quite widespread amongst hunters. For example, Adams reports an online comment from a hunter whose grandfather used to say "the hunt is more important than the kill" (internet comment, in Adams 2013, 50). Even so, the intent to kill often remains, regardless of success, and is important to the desired experience. Franklin asserts that non-consumptive forms of wildlife leisure "derive from relations that create distance rather than proximity, separation rather than interaction and spectacle rather than sensual, embodied relations" (Franklin 2008, 36).

In contrast, there are increasing numbers of hunters who are primarily driven by the acquisition of meat (Cerulli 2010a). Some authors claim that self-provisioning hunters, for whom the goal is supplementing food stores, rather than the sole source (Curnutt 1996), are a type of recreational hunter, as their practices are not necessary for survival. Michael Adams describes his hunting for self-provisioning as "opposed to 'subsistence', because we don't need to do this, we choose to do it" (Adams 2013, 49). However, the recreation–subsistence demarcation is not firm. Both types of hunters share many of the material and psychological benefits of hunting (Cerulli 2010a; List 2004). Subsistence hunters often enjoy the hunt, and there is strong consensus amongst both hunters and non-hunters that recreational hunters should eat the meat from the animals that they kill (Cohen 2014; Fischer *et al.* 2013).

Why might hunted meat be more 'ethical' than farmed meat?

Recent decades have seen a surge in attention to food ethics. Popular books such as Michael Pollan's *The Omnivore's Dilemma* and films including *Forks over Knives, Cowspiracy* and *Food Inc.* have made such discourses commonplace. In addition, the internet and magazines have provided new sources of information about alternative food movements (Teitelbaum and Beckley 2006). Coupled with the socio-economic prosperity in post-industrial countries, consumers now have the ability to be more discerning in their food choices (Manfredo, Teel and Bright 2003). There are two consequences of these trends relevant for our analysis: first, criticisms increasingly have been directed at conventional meat production methods, especially in relation to ethics, health and the environment (Gressier 2016), including their contribution to greenhouse gas emissions and climate change (e.g., Bauer and English 2011; Garnaut 2008); second, anti-neoliberal movements such as locavorism and the slow food movement, which represent rejections of the assumptions underlying modern food production and distribution such as low cost and convenience (Guthman 2008; Leroy and Degreef 2015). A large focus in these movements is 'self-provisioning,' producing material goods that are consumed or shared, but not sold (Teitelbaum and Beckley 2006); as a result, hunting has come to be viewed by some as an alternative means of sourcing meat.

It is not only alternative food movements that are promoting game consumption. Food writers and celebrity chefs – "modern food heroes" – are playing a role in this new promotion of game by "embracing pre-industrial values" (Dubecki 2013). Drivers range from blogs and local newspapers (Dubecki and Han 2013) to television programmes such as "Masterchef" (Gressier 2016, 50) and promotion by globally recognised figures such as Jamie Oliver (Lunney 2012, 13; see also Phillipov 2016 for further discussion of television cooking shows and food politics). For some segments of society, "gastronomic novelty" is a significant motivator for growing interest in wild meat consumption (Gressier 2016, 59). In some locales, game is a delicacy due to how difficult it can be to acquire.

In Europe and the US, recent increases in demands for game meat stem from consumer preferences for healthier and more ethical meat sources (Hoffman and Wiklund 2006). In a recent survey in New York State, 88% of respondents stated that their primary motivations to hunt were to harvest local, natural meat (Quartuch *et al.* 2016). As Cerulli asks, "where else can you get organic, free-range, grass-fed, arguably cruelty-free, roughly 100-mile diet?" (Cerulli 2010b, 3). A trend away from intensively farmed meat has been noted in Australia where wild-caught game is seen by some members of the public as a more ethical alternative (Gressier 2016), a view echoed in philosophical literature (Bruckner 2007). "Within this ideology, hunting meat is seen as more noble than purchasing it, while wild meat is seen as preferable to farmed" (Gressier 2016, 58). Meat is one of the products which consumers desire the most to be natural (Rozin *et al.* 2004) and hunted meat has been found to be considered more deserving of adjectives such as 'good' or 'green' than intensively farmed meat (Fischer *et al.* 2013). This type of discourse is especially relevant in Australia, given that Western-style agriculture has only been utilised relatively recently and hence is considered by some to be 'unnatural' (Saltzman, Head and Stenseke 2011).

A review of the scholarly literature reveals two main perspectives on hunting as compared to animal farming. First, deaths of animals from hunting are morally superior to those raised for meat and killed in slaughterhouses (Bruckner 2007; see also Bauer and English 2011 for discussion of problems with modern meat production). An Australian survey found 51% agree or strongly agree that 'factory-farming' methods are unnatural, and 52% that they are cruel (Franklin 2007). This contrast with 'factory-farms' is a common theme in hunting literature: "What is worse? Hunting? Or the bio-industry where pigs who never saw daylight are being fully automatically butchered?" (Koelewijn 2014, in Van Heijgen 2015, 45). Or as an internet post, quoted in Adams (2013, 50) puts it:

> Hunting? A creature is peacefully in its own domain, it is shot. How is that worse than being carried for hours in a truck, being forced into a crush, hearing the bellows of other creatures, being physically restrained at the peak of terror, then culled? Or bred in an area hardly big enough for the creature to move, then bundled off to be slaughtered en masse?

A second perspective is that the problem of animal death cannot be avoided even by abstaining from meat consumption, and thus hunting is preferable. Adams writes that:

> it is not really possible to avoid the deaths of animals in human lives, although you can distance yourself at various scales from those deaths. At its best, hunting can do the opposite: you do the killing yourself, taking personal moral responsibility.
>
> (Adams 2013, 48)

Because even plant food production requires the killing of pests, loss of habitat for wild animals, and accidental deaths due to farm machinery, some hunters see killing a smaller number of larger animals as an ethically preferable alternative (Cerulli 2010b; Davis 2003). Hunting is also seen as a way to confront the nature of these necessary deaths. Thus, killing one's own meat is "being reconfigured as honourable engagements within an albeit unpleasant reality" (Gressier 2016, 58), in that it makes "the materiality of food production explicit" (Peterson *et al.* 2010, 127). If you are a hunter, "You can no longer go to the supermarket to buy meat and ignore the consequences", states Rohan Anderson, who moved out of Melbourne and now only eats meat he raises or hunts himself (Dupleix 2012). For some hunters, involving their children in hunting activities provides an opportunity to teach ethical meat consumption, as described by a participant in Bray et al (2016, 6) "[The children] ... understand the implications of taking a life to feed a life".

Hunting in Australia

Finding exact numbers of hunters in Australia is difficult, but a recent survey estimates that there are 200,000–350,000 recreational hunters in Australia, or 1.5% of the population (Finch *et al.* 2014), which may be linked to the highly limited amount of land where recreational hunting is permitted compared to other countries (Woods and Kerr 2010; Burgin 2015). Even with low hunter numbers compared to Scandinavia and North America, there are over 50 recreational hunting clubs in Australia, most of which are based in the states of Victoria and New South Wales (Craig-Smith and Dryden 2008). Little empirical research has been done on hunting in Australia compared to scholarship in Europe and North America (Fitzgerald, Fitzgerald and Davidson 2007). Of this research, the majority is concerned with the hunting practices of Indigenous Australian people. As the relationships between humans and animals vary "according to historical, regional and cultural contingencies" (Franklin 1996, 43), it is impossible to understand the place of hunting in Australia, either for food or leisure, without understanding the contingencies that make Australia unique. These circumstances also mean that much of the research done on hunting in Europe and North America may not be directly applicable to the Australian context.

Animals in Australia

One critical issue is the place of animals in the history of Australian colonisation and the impact that this legacy has had on today's attitudes towards hunting in the country. Australia was colonised for its value as a penal colony, not for the economic value of its land or wildlife, both of which were unfamiliar to Europeans (Burgin 2015). Thus, early settlers often did not appreciate native animals either economically or recreationally (Franklin 2011; Smith 2011), although there is increasing evidence that some native animals were in fact consumed in the early colony (Newling 2011; Santich 2012; Cushing 2016). As time went on, native animals came to be seen as food for the poor (Bauer and English 2011) and agricultural pests. The small number of native species that were suitable for sport were soon hunted to extinction (Franklin 1996). As a result, species such as foxes, rabbits and deer were introduced by the colonial upper classes for private hunting (Burgin 2015; Franklin 1996).

In the decades preceding Federation in 1901, attitudes towards native animals underwent a marked shift. In an attempt to create a national identity, native species were valorised as part of an 'Australianisation' discourse (Franklin 1996, 2008; Gressier 2016). In the early part of the 20th century, particularly during the Depression, introduced species such as rabbit became the "poor-man's meat" (Gressier 2016, 50) and the public rallied around campaigns to prevent hunting of possums and koalas for skins, creating an anti-hunting sentiment that has lasted in Australia until the present day (Burgin 2015). Despite this complex history, together with Australia's status as one of the most urbanised countries in the world, hunting tends to be associated with "a formative origin myth in Australian culture" (Franklin 1996, 54), one that valorises the outback and the farmer who lives off the land. Hunting culture in Australia is not as focused on education as it is in Europe and the United States, and has relatively little contact with international hunting organisations (Bauer and English 2011).

An additional issue arising out of Australia's colonial history is the dichotomy between native and introduced species. All game species (particularly deer and pigs) in Australia are introduced species (Bauer and English 2011; Craig-Smith and Dryden 2008). Native species are much more valued than introduced species, and thus their hunting is far more controversial (Fitzgerald, Fitzgerald and Davidson 2007; Franklin 1996; Franklin 2007). Native species are often rejected precisely because they are closely identified with the nation. One respondent interviewed by Waitt and Appleby expressed disgust at the prospect of eating kangaroo meat: "I'm an Aussie and I don't think . . . I wouldn't eat my national emblem" (Waitt and Appleby 2014, 95). An additional issue is the distinction made between 'useful' introduced species and 'ferals,' inasmuch as cows, sheep and other farm-related animals provoke different reactions. This discrepancy may be due, at least in part, to the way that domestic working animals factor into the creation myth of Australian identity and form a central part of the economy (Gressier 2016).

Once introduced species have become well-established in local ecosystems, doubt can be created in the public consciousness even about their introduced status. For instance, surveyed landholders in Queensland (Finch and Baxter 2007) associated deer more with native species (50% agree, 40% disagree) than feral pests (39% agree, 51% disagree), believed that wild deer populations should be preserved for future generations to enjoy (56% agree, 32% disagree), and were not considered 'significant' pests in comparison to wild cats, pigs and rabbits. These perceptions may be related to the history of deer legislation in Queensland where deer were protected fauna from 1952 to 1992, giving them a legal status similar to that accorded to native species (Finch and Baxter 2007). In contrast, only 19% of respondents in an older Victorian survey believed that "introduced animals should be considered to be 'native' if their populations were established for 100 years or more" (Johnston and Marks 1997, in Fitzgerald, Fitzgerald and Davidson 2007, 14). Some authors suggest that native species, such as kangaroo, should be redefined as game for environmental and economic reasons (Bauer and English 2011), and hence not protected by legislation covering native species.

Motivations for hunting in Australia

Pest control is the major self-reported motivation of Australian hunters (Finch *et al.* 2014). Thus, unlike Britain and the United States, hunting in Australia is seen more as work than leisure or pleasure (Franklin 1996). A study of wild-pig management found that "benefits from the 'wild boar' meat export industry and recreational hunting are attractive and appear to be a factor in people's control/ management preferences for feral pigs" (Fitzgerald, Fitzgerald and Davidson 2007, 32). However, with the exception of pigs, hunting does not constitute a major part of the management for introduced species (Franklin 2011), although recreational hunting and game meat harvesting were the favoured techniques to manage deer populations among surveyed landholders in Queensland (Finch and Baxter 2007). As the species recreationally hunted in Australia are all introduced, a large part of the justification for hunting is that hunters are serving a vital ecological function by preventing overpopulation. This argument is very common in the hunting literature (Bauer and English 2011; Enck, Decker and Brown 2000; Heberlein 1991; Oldfield 2015; Peterson 2004; Vitali 1990), and also contributes towards the mitigation of negative views towards hunters, as they are both "killers of European interlopers [and] at the same time heroic defenders of a fragile Australian natural purity" (Franklin 1996, 52). Although recent data is lacking, a survey conducted in 2000 found 68% of Australians agree or strongly agree that it is acceptable to hunt feral species that degrade the environment (Franklin 2007). Furthermore, some contend that objections to hunting based on the potential damage it can create for the ecosystem are not as relevant in Australia as elsewhere, as all hunted species are introduced (Vitali 1990). However, there is tension between pest management efforts and maintaining populations for recreational hunters. Hunting advocacy groups have been successful at lobbying state governments, especially in Victoria, to create management policy which serves the interests of hunters, but which has been noted to be ineffective at reducing deer numbers (Barber 2016; Bilney 2013; Burgin *et al.* 2015; Shoebridge and Hopley 2014).

A final rationale relevant to hunting in Australia relates to waste and food security, as large swathes of the continent are unsuitable for farming. Hunted animals are a useful part of the food production network by converting inedible protein, such as grass, into edible meat (Adams 2013). As feral animal populations are so large, they could easily be used as a renewable meat source. Bauer and English estimate that up to 10 million wild pigs could be harvested for meat each year without endangering the population (2011). Even if we do not hunt feral animals specifically to eat in Australia, meat from culled animals could still be consumed domestically or exported to avoid wasting edible meat (Houghton 2014; Lagan 2016), so long as they were killed in a manner that would allow the meat to be usable.

Attitudes to hunting in Australia

There is evidence that public attention to hunting is growing, but that public opinion is becoming more critical of hunting; however, recent Australian empirical data on this issue is scant. Although we argue throughout this chapter that the context for hunting in Australia differs from that of Europe and North America, studies in these locales show that attitudes differ according to the purpose and target species, as well as other variables such as rurality, gender and game meat consumption (Gamborg and Jensen 2017; Ljung *et al.* 2012) . A 2000 survey found that 14% of Australians would be less likely to hunt than "a few years ago" and only 2% were more likely to hunt (Franklin 2007, 21). Many people are more likely to modify their activities to be more 'animal-friendly,' such as eating less meat, buying free-range eggs, avoiding animal-tested products and donating to animal charities than previously (Franklin 2007). A content analysis of coverage by the Tasmanian newspaper *The Mercury* in the second half of the 20th century also shows "rises in zoocentrism and sentimentality … [and] attention to native animals and … decline in stories on angling and hunting" (Franklin and White 2001, 235).

Hunting remains a controversial issue. Hunters "must convince their critics that deer hunting is not cruel, that hunters are not blood-thirsty killers but are responsible citizens who respect the deer and other wildlife" (Harrison and Slee 1995, iv). The question of the social acceptability of hunting is made easier in Australia insofar as the primary game species are introduced, and the discourse of native versus non-native allows hunters to claim they are serving a vital ecological function. Some also point out that recent legislation and regulation of hunting in Australia (e.g., the creation of the New South Wales Game Council) have made processes more transparent and contributed to legitimising hunting in the public consciousness (Bauer and English 2011). Additionally, hunting for food is far less controversial, whereas hunters who do not eat the animals they kill are widely condemned (Fischer *et al.* 2013). These factors combine in opposition to hunting activities such as the Victorian duck hunts, which not only kill native species, but result in many birds being discarded rather than consumed (Munro 1997). A 2013 editorial in Australian newspaper *The Age* condemned duck hunting as serving "no real purpose other than to stir the adrenalin of gum-booted shooters" ("Failure on so many levels" 2013).

Another important factor shaping Australian attitudes toward hunting relates to perceptions of gun ownership, which became more negative following the 1996 Port Arthur massacre and the subsequent amnesties and buybacks of guns together with tightening of Australia's state and territory gun laws, which are now among the strictest worldwide (Alpers 2013; Chapman 2013). Australians generally have held more negative images of hunters than Europeans and Americans (Bauer and English 2011; Burgin 2015). Hunting culture is linked to Anglo-Saxon traditions of small numbers of hunters who follow a 'sporting' code and hunt on private land, but is also strongly related to pest control (Sharp and Wollscheid 2009). Despite this history, hunting more generally in Australia is perceived as a middle- or lower-class pursuit (Franklin 1996), often associated with 'bogans' (a slang term for the lower class (Adams 2013)) or recent migrants. Goat hunting, for example, is especially popular with hunters of "southern European and Middle Eastern backgrounds" (Bauer and English 2011, 113) due to the popularity of goat meat in their traditional cuisines.

Gun culture is a matter of some debate in Australia. For instance in the *Bateman's Bay Post* in response to a local hunting expo, defenders of hunting claimed that "contrary to popular publicity, the 'gun culture' is tolerant, responsible and respectful of others and the law. Shooters are targets for prejudice and vilification because 'guns are bad'" (Burg 2013). On the other hand, critics ask, "how responsible is it to encourage children to think killing animals is fun?" (Cruttenden 2013). Wildlife management in Australia has more of a focus on being 'humane' than is the case in the United States (Fitzgerald, Fitzgerald and Davidson 2007), with the level of community resistance to management techniques increasing with lethality (Burgin *et al.* 2015). Thus, some have concluded that Australia has more recently become a "bastion" of animal-rights groups (Bauer and English 2011, 228). This is a change even from the 1990s when Franklin observed that in Australia "anti-hunting activity is discernibly low key" (1996, 52).

The 'right' way to hunt? Eating as a justification for recreational hunting

At the intersection of these discourses, we are confronted with an overarching and fundamental question: is there a 'correct' way to hunt? Arthur Bentley, the founder of the Australian Deer Research Foundation, a hunters' advocacy group, acknowledged that a certain level of professionalism was required of hunters "in today's climate of protest and confrontation" (Harrison and Slee 1995, iii). Hunting involves "an ethical code, which the hunter formulates for himself, and must live up to without the moral support of bystanders" (Leopold 1933, in Bauer and English 2011, 5). Most hunters believe in principles such as that of a 'fair chase' (Sharp and Wollscheid 2009). Brian Luke (1997) provides an overview of the points shared by various ethical hunting codes: safety first, obey the law, give fair chase, harvest the game, aim for quick kills and retrieve the wounded.

Against this backdrop we return to the distinction between hunting to kill and killing to hunt, and to the questions of whether the eating of game meat is merely a justification for recreational hunting and where the line between recreation and self-provisioning is. These are important questions to ask if we wish to understand self-provisioning as a component of use of animals in leisure. Even subsistence hunters, who must hunt to survive, derive (and prize) social, cultural and psychological benefits (Emery and Pierce 2005) from their hunting activities. While some hunters hunt primarily for food, for others the experience of eating wild-caught game is the most significant outcome (Teitelbaum and Beckley 2006). A study in Denmark found that the vast majority of participants hunted to experience nature and be with friends, with 90% saying they did not hunt for meat (Hansen, Peterson and Jensen 2012). A literature review from a variety of countries concluded that while acquisition of meat was a greater motivation than trophy hunting, there were many other reasons people hunted, with experiencing nature and socialising with other hunters being the most important (Woods and Kerr 2010). An Australian study found that the primary motivation to hunt was pest control, followed closely by recreation and then meat (Finch *et al.* 2014). Critics of hunting have observed what they view as the 'hypocritical' nature of some recreational hunters who use discourses of sustainability, environmentalism, or spiritual identification with nature "to camouflage and to legitimate violence and biocide" (Kheel 1995, 87). Even hunters who claim their primary motivation to be the acquisition of meat are aware that they also appreciate other aspects of the experience, as can be observed in statements such as "I enjoy the thrill of the chase and the feeling of satisfaction after a successful hunt" (Helen, a hunter, in Merskin 2010, 229–230). The experience becomes embedded in understandings of hunting:

> [Y]ou know, certainly, the fish, or, or the venison or whatever happens to be, is the, the end result, but certainly the whole package is actually getting to that end result ... you know, organizing the trip, going on the trip, the boat ride out there, the company, the company on the trip ...
>
> (Marty, a hunter, in Reis 2009, 581)

There is a strong feeling that eating meat legitimises the killing of animals, regardless of the primary motivation: "if you want to shoot a buck and have it mounted, that's fine. Do something with the meat. If you want to kill 10 does, that's fine too, as long as you do something with the meat" (Ed, a hunter, in Littlefield and Ozanne 2011, 346). Waste, then, becomes a primary concern: "a common denominator of all sporting codes is not to waste good meat" (Leopold 1966, in Simpson and Cain 2000, 189–190). For example, The Australian Deer Association (2014), a recreational hunting organisation, states in its code of conduct that "If a deer is shot, the whole carcass should be taken, but if this is not possible the venison should be utilised".

Conclusion

In this chapter, we have synthesised scholarly literature, media articles and governmental agency reports to explore the motivations underlying hunting among Australians, with a view to examining whether self-provisioning activities are consistent with popular ideas of ethical consumption. This exploration in turn allows reflections on when and whether hunting should be viewed as a leisure activity. Our findings reveal the complexities associated with hunting practices, including views about how it can be viewed as a form of subsistence and provisioning, and thus is morally permissible as it is not simply or primarily a leisure activity. This debate suggests that further consideration is warranted for other activities that are 'hybrid' in similar ways, including what the implications are for our understandings of them as part of leisure. The case of hunting also raises themes that could fruitfully be considered through the lens of 'dark leisure' (Stone 2013), especially as it clearly raises questions about morality and taboos associated with leisure activities involving death.

We have highlighted that both the discourse around which animals should or should not be hunted because of their place in the Australian landscape and public views towards gun ownership provide a unique context for discussions about meat obtained by hunting and wild-harvesting as types of ethical consumption which invite further empirical exploration. As increasing attention to animal welfare and animal rights in association with hunting practices may affect public attitudes, the issues raised in this chapter will need to continue to be monitored. This chapter also highlights the need for deeper understanding of the values and motivations associated with hunting practices given their evolution and diversity in order to facilitate more constructive dialogues about the role of hunting in food production and consumption in Australia and beyond.

References

Adams, Michael. 2013. "Redneck, barbaric, cashed up bogan? I don't think so": Hunting and Nature in Australia. *Environmental Humanities* 2 (1):43–56. doi: 10.1215/22011919-3610342.

Alpers, Philip. 2013. The big melt: How one democracy changed after scrapping a third of its firearms. In *Reducing Gun Violence in America: Informing Policy with Evidence and Analysis*, edited by Daniel W. Webster and Jon S. Vernick, 205–211. Baltimore: Johns Hopkins University Press.

Australian Deer Association 2014. Code of Conduct 2014. Accessed 20 January. www.austdeer.com.au/about-ada/code-of-conduct/.

Barber, Greg. 2016. As deer control fails, time to call in professionals. *The Weekly Times*, 14 September.

Bauer, Johannes, and Anthony English. 2011. *Conservation Through Hunting: An Environmental Paradigm Change in NSW*. Orange, NSW: Game Council NSW.

Bilney, Rohan. 2013. The protected pest: deer in Australia. *The Conversation*, 11 February. Accessed 3 December 2016. http://theconversation.com/the-protected-pest-deer-in-australia-11452.

Bray, Heather J., Sofia C. Zambrano, Anna Chur-Hansen and Rachel A. Ankeny. 2016. Not appropriate dinner table conversation? Talking to children about meat production. *Appetite* 100:1–9. doi: http://dx.doi.org/10.1016/j.appet.2016.01.029.

Bruckner, Donald W. 2007. Considerations on the morality of meat consumption: hunted-game versus farm-raised animals. *Journal of Social Philosophy* 38 (2):311–330. doi: 10.1111/j.1467-9833.2007.00381.x.

Burg, Oren. 2013. Straight shooting on 'gun culture'. *Bateman's Bay Post*, 24 July.

Burgin, Shelley. 2015. Why the difference in the recreational hunting ethic between Australians and North Americans? An opinion with emphasis on "furbearers". *International Journal of Environmental Studies* 72 (5):770–783. doi: 10.1080/00207233.2015.1077592.

Burgin, Shelley, Mariama Mattila, Daryl McPhee and Tor Hundloe. 2015. Feral deer in the suburbs: an emerging issue for Australia? *Human Dimensions of Wildlife* 20 (1):65–80. doi: 10.1080/10871209.2015.953274.

Cerulli, Tovar. 2010a. Hunting and heresy: A skirmish with Ortega y Gasset. *Tovar Cerulli: Catalyzing Insights for Conservation*, June 14. http://tovarcerulli.com/hunting-and-heresy/.

Cerulli, Tovar. 2010b. Hunting like a vegetarian: same ethics, different flavor. In *Hunting: In Search of the Wild Life*, edited by Nathan Kowalsky, 45–55. Chichester, West Sussex: Wiley-Blackwell.

Chapman, Simon. 2013. *Over Our Dead Bodies: Port Arthur and Australia's Fight for Gun Control*. Sydney: Sydney University Press.

Cohen, Erik. 2014. Recreational hunting: ethics, experiences and commoditization. *Tourism Recreation Research* 39 (1):3–17. doi: 10.1080/02508281.2014.11081323.

Craig-Smith, Stephen J., and Gordon McL. Dryden. 2008. Australia as a safari hunting destination for exotic animals. In *Tourism and the Consumption of Wildlife: Hunting, Shooting and Sport Fishing*, edited by Brent Lovelock, 268–280. London: Routledge.

Cruttenden, Susan. 2013. Say no to promotion of guns. *Bateman's Bay Post*, 28 November.

Curnutt, Jordan. 1996. How to argue for and against sport hunting. *Journal of Social Philosophy* 27 (2):65–89. doi: 10.1111/j.1467-9833.1996.tb00238.x.

Cushing, Nancy E. 2016. Interspecies entanglements of eating kangaroo, 1788–1850. *History Australia*, 13 (2):286–299.

Davis, Steven L. 2003. The least harm principle may require that humans consume a diet containing large herbivores, not a vegan diet. *Journal of Agricultural and Environmental Ethics* 16 (4):387–394. doi: 10.1023/a:1025638030686.

Dubecki, Larissa. 2013. The happy hunting grounds of the modern food hero. *The Age*, 3 March.

Dubecki, Larissa, and Esther Han. 2013. Hunter-gatherer master chefs. *The Sydney Morning Herald*, 3 March.

Dupleix, Jill. 2012. Future set to dish up the past. *The Age*, 17 January.

Emery, Maria R., and Alan R. Pierce. 2005. Interrupting the telos: locating subsistence in contemporary US forests. *Environment and Planning A* 37 (6):981–993

Enck, Jody W., Daniel J. Decker and Tommy L. Brown. 2000. Status of hunter recruitment and retention in the United States. *Wildlife Society Bulletin (1973–2006)* 28 (4):817–824.

Failure on so many levels. 2013. *The Age*, 14 May.

Finch, N. A., and G. S. Baxter. 2007. Oh deer, what can the matter be? Landholder attitudes to deer management in Queensland. *Wildlife Research* 34 (3):211–217. doi: http://dx.doi.org/10.1071/WR06002.

Finch, Neal, Peter Murray, Julia Hoy and Greg Baxter. 2014. Expenditure and motivation of Australian recreational hunters. *Wildlife Research* 41 (1):76–83.

Fischer, Anke, Vesna Kereži, Beatriz Arroyo, Miguel Mateos-Delibes, Degu Tadie, Asanterabi Lowassa, Olve Krange and Ketil Skogen. 2013. (De)legitimising hunting – discourses over the morality of hunting in Europe and eastern Africa. *Land Use Policy* 32:261–270. doi: http://dx.doi.org/10.1016/j.landusepol.2012.11.002.

Fitzgerald, Gerard, Nic Fitzgerald and Carl Davidson. 2007. *Public Attitudes Towards Invasive Animals and Their Impacts*. Invasive Animals Cooperative Research Centre, Canberra.

Franklin, Adrian. 1996. Australian hunting and angling sports and the changing nature of human-animal relations in Australia. *Journal of Sociology* 32 (3):39–56. doi: 10.1177/144078339603200303.

Franklin, Adrian. 2007. Human-nonhuman animal relationships in Australia: an overview of results from the first national survey and follow-up case studies 2000–2004. *Society & Animals* 15 (1):7–27. doi: doi:http://dx.doi.org/10.1163/156853007X169315.

Franklin, Adrian. 2008. The 'animal question' and the 'consumption' of wildlife. In *Tourism and the Consumption of Wildlife: Hunting, Shooting and Sport Fishing*, edited by Brent Lovelock, 31–44. London: Routledge.

Franklin, Adrian. 2011. An improper nature? Introduced animals and 'species cleansing' in Australia. In *Human and Other Animals: Critical Perspectives*, edited by Bob Carter and Nickie Charles, 195–216. London: Palgrave Macmillan.

Franklin, Adrian, and Robert White. 2001. Animals and modernity: changing human–animal relations, 1949–98. *Journal of Sociology* 37 (3):219–238. doi: doi:10.1177/144078301128756319.

Gamborg, Christian, and Frank Søndergaard Jensen. 2017. Attitudes towards recreational hunting: A quantitative survey of the general public in Denmark. *Journal of Outdoor Recreation and Tourism* 17:20–28.

Garnaut, Ross. 2008. *The Garnaut Climate Change Review*. Melbourne: Cambridge University Press.

Gressier, Catie. 2016. Going feral: wild meat consumption and the uncanny in Melbourne, Australia. *The Australian Journal of Anthropology* 27 (1):49–65. doi: 10.1111/taja.12141.

Guthman, Julie. 2008. Neoliberalism and the making of food politics in California. *Geoforum* 39 (3):1171–1183.

Hansen, Hans Peter, M. Nils Peterson and Charlotte Jensen. 2012. Demographic transition among hunters: a temporal analysis of hunter recruitment dedication and motives in Denmark. *Wildlife Research* 39 (5):446–451.

Harrison, Mike, and Ken Slee. 1995. *The Australian Deerhunter's Handbook*. 2nd ed. Melbourne: Australian Deer Research Foundation.

Heberlein, Thomas A. 1991. Changing attitudes and funding for wildlife: preserving the sport hunter. *Wildlife Society Bulletin (1973–2006)* 19 (4):528–534.

Hoffman, L. C., and E. Wiklund. 2006. Game and venison – meat for the modern consumer. *Meat Science* 74 (1):197–208. doi: http://dx.doi.org/10.1016/j.meatsci.2006.04.005.

Houghton, Des. 2014. How to combat ferals: eat em. *Courier-Mail*, 28 June.

Kheel, Marti. 1995. License to kill: an ecofeminist critique of hunters discourse. In *Animals and Women: Feminist Theoretical Explanations*, edited by C. Adams and J. Donovan, 85–125. Durham, NC: Duke University Press.

King, Roger J. H. 2010. Hunting: a return to nature? In *Hunting: In Search of the Wild Life*, edited by Nathan Kowalsky, 149–160. Chichester, West Sussex: Wiley-Blackwell.

Kover, T. R. 2010. Flesh, death, and tofu: hunters, vegetarians, and carnal knowledge. In *Hunting: In Search of the Wild Life*, edited by Nathan Kowalsky, 171–183. Chichester, West Sussex: Wiley-Blackwell.

Lagan, Bernard. 2016. Send wild dogs for the Chinese to eat, urges ecology expert. *The Times*, 17 August.

Leader-Williams, Nigel. 2009. Conservation and hunting: friends or foes? In *Recreational Hunting, Conservation and Rural Livelihoods*, edited by Barney Dickson, John Hutton and William M. Adams, 9–24. Oxford: Wiley.

Leroy, Frédéric, and Filip Degreef. 2015. Convenient meat and meat products. Societal and technological issues. *Appetite* 94:40–46. doi: http://dx.doi.org/10.1016/j.appet.2015.01.022.

List, Charles J. 2004. On the moral distinctiveness of sport hunting. *Environmental Ethics* 26 (2):155–169.

Ljung, Per E., Shawn J. Riley, Thomas A. Heberlein and Göran Ericsson. 2012. Eat prey and love: game-meat consumption and attitudes toward hunting. *Wildlife Society Bulletin* 36 (4):669–675.

Littlefield, Jon, and Julie L. Ozanne. 2011. Socialization into consumer culture: hunters learning to be men. *Consumption Markets & Culture* 14 (4):333–360. doi: 10.1080/10253866.2011.604494.

Luke, Brian. 1997. A critical analysis of hunters' ethics. *Environmental Ethics* 19 (1):25–44.

Lunney, Daniel. 2012. Wildlife management and the debate on the ethics of animal use. I. Decisions within a State wildlife agency. *Pacific Conservation Biology* 18 (1):5–21.

Manfredo, Michael, Tara Teel and Alan Bright. 2003. Why are public values toward wildlife changing? *Human Dimensions of Wildlife* 8 (4):287–306. doi: 10.1080/716100425.

Marx, Jack. 2012. A Christian shooting party. *The Sydney Morning Herald*, 8 September.

Merskin, Debra. 2010. The new Artemis? Women who hunt. In *Hunting: In Search of the Wild Life*, edited by Nathan Kowalsky, 225–238. Chichester, West Sussex: Wiley-Blackwell.

Munro, Lyle. 1997. Framing cruelty: the construction of duck shooting as a social problem. *Society & Animals* 5 (2):137–154. doi: doi:https://doi.org/10.1163/156853097X00042.

Munro, Peter. 2012. Young, old and angry gather for the duck shoot. *The Age*, 18 March.

Newling, Jacqui. 2011. Dining with strangeness: European foodways on the Eora frontier. *Journal of Australian Colonial History* 13:27–48.

Oldfield, Jamie-Lee. 2015. Shooters firm on hunting as viable control. *The Land*, 19 February.

Peterson, M. Nils. 2004. An approach for demonstrating the social legitimacy of hunting. *Wildlife Society Bulletin (1973–2006)* 32 (2):310–321.

Peterson, M. Nils, Hans Peter Hansen, Markus J Peterson and Tarla R. Peterson. 2010. How hunting strengthens social awareness of coupled human-natural systems. *Wildlife Biology in Practice* 6 (2):127–143.

Phillipov, Michelle. 2016. The new politics of food: television and the media/food industries. *Media International Australia*, 158 (1):90–98.

Pollan, Michael. 2006. *The Omnivore's Dilemma: A Natural History of Four Meals*. New York: Penguin Press.

Quartuch, Michael R., Richard C. Stedman, Daniel J. Decker, William F. Siemer, Meghan S. Baumer and Lincoln R. Larson. 2016. *Taking a Non-traditional Path to Hunting in New York: Insights and Implications for Recruitment and Retention*. Ithaca, NY: Cornell University.

Reis, Arianne Carvalhedo. 2009. More than the kill: hunters' relationships with landscape and prey. *Current Issues in Tourism* 12 (5–6):573–587. doi: 10.1080/13683500903042881.

Rozin, Paul, Mark Spranca, Zeev Krieger, Ruth Neuhaus, Darlene Surillo, Amy Swerdlin and Katherine Wood. 2004. Preference for natural: instrumental and ideational/moral motivations, and the contrast between foods and medicines. *Appetite* 43 (2):147–154. doi: http://dx.doi.org/10.1016/j.appet.2004.03.005.

Saltzman, Katarina, Lesley Head and Marie Stenseke. 2011. Do cows belong in nature? The cultural basis of agriculture in Sweden and Australia. *Journal of Rural Studies* 27 (1):54–62. doi: http://dx.doi.org/10.1016/j.jrurstud.2010.09.001.

Santich, Barbara. 2012. *Bold Palates: Australia's Gastronomic Heritage*. Adelaide: Wakefield Press.

Sharp, Robin, and Kai-Uwe Wollscheid. 2009. An overview of recreational hunting in North America, Europe and Australia. In *Recreational Hunting, Conservation and Rural Livelihoods*, edited by Barney Dickson, John Hutton and William Adams, 23–35. Oxford: Wiley.

Shoebridge, David, and Caroline Hopley. 2014. Amateur hunting: It's a blood-sport not a conservation measure. *Nature New South Wales* 58 (3):28–29.

Simpson, Steven V., and Kelly D. Cain. 2000. Recreation's role in the environmental ethics dialogue: the case of Aldo Leopold and the morality of hunting. *Leisure/Loisir* 25 (3–4):181–197. doi: 10.1080/14927713.2000.9649916.

Smith, Nicholas. 2011. Blood and soil: nature, native and nation in the Australian imaginary. *Journal of Australian Studies* 35 (1):1–18. doi: 10.1080/14443058.2010.541475.

Stone, Philip R., and Richard Sharpley 2013. Deviance, dark tourism and "dark leisure": towards a (re)configuration of morality and the taboo in secular society. In *Contemporary Perspectives in Leisure: Meanings, Motives and Lifelong Learning*, edited by Sam Elkington and Sean Gammon, Abingdon: Routledge, pp. 54–64.

Teitelbaum, Sara, and Thomas Beckley. 2006. Harvested, hunted and home grown: the prevalence of self-provisioning in rural Canada. *Journal of Rural and Community Development* 1 (1):114–130.

Van Heijgen, Eugenie J. 2015. *How the hunter became the hunted: A discourse analysis on how hunters cope with the emerged resistance against hunting practices in the Netherlands*. Master of Landscape Architecture and Planning, Department of Environmental Sciences, Wageningen University.

Vitali, Theodore R. 1990. Sport hunting: moral or immoral? *Environmental Ethics* 12 (1):69–82.

Wade, Maurice L. 1990. Animal liberationism, ecocentrism, and the morality of sport hunting. *Journal of the Philosophy of Sport* 17 (1):15–27.

Waitt, G., and Bryce Appleby. 2014. "It smells disgusting": plating up kangaroo for a changing climate. *Continuum* 28 (1):88–100. doi: 10.1080/10304312.2013.854863.

Woods, Amelia, and Geoffrey N. Kerr. 2010. *Recreational Game Hunting: Motivations, Satisfactions and Participation*. Lincoln University, Canterbury, New Zealand.

13 Conclusions

Charting a way forward

Neil Carr and Janette Young

Introduction

Part of the joy and fear in constructing an edited book is that after building the rationale for it and forwarding a call for submissions to the world, you are never sure what, if anything, you will receive. Stepping into the unknown (will anybody respond and if they do will the proposed chapters be publishable) is what generates the fear. Yet the same unknown provides opportunity. It allows us collectively to reach out to people and areas of study that are new. In this way, an edited book can provide an environment for academic interactions. It can provide fertile ground for the development of thinking on new and relatively unexplored subjects.

Having seen many people respond to the call for chapters for this book and its twin, which focuses on domestic/domesticated animals (Young & Carr, forthcoming, a), and subsequently produce high-quality cutting-edge work, our fears have been allayed. What has been produced is a group of chapters that collectively push forward understandings of the position of animals in human leisure and the leisure needs and desires of animals. For us, the editors, the result is both a recognition of the vibrant state of studies of animals and leisure, and a new awareness of the areas of work yet to be undertaken to help ensure the welfare of animals in human leisure, and their welfare through their own leisure.

What follows in this chapter is a reflection on the content of the book and the highlighting of some emergent issues. This is followed by some thoughts about how to move studies of animals and leisure forward.

Reflections and emergent issues: a welfarist focus

It is clear throughout the chapters of this book that animal welfare (including their access to leisure) is a significant concern of humans within the leisure experience. However, it is equally clear that animals, even those living in the wild, are at least to some extent beholden to humans to ensure this welfare is in place. This is a recognition that humans have power over animals; that the world is dominated by humans. The question, as noted by Curtin and Green in Chapter 4 about zoo visitors' ethical perspectives regarding the keeping of wild animals in captivity, is what happens when human desires come up against the welfare needs of animals. Ultimately, animals are dependent on the moral compass of humans, individually and collectively. Hopefully this compass will point people towards their obligations to animals rather than allowing their personal desires to take priority. However, this is an unsatisfactory situation for proponents of the animal rights position.

It is clear that this is an animal welfarist book and not one that aligns with animal rights proponents. This is arguably no surprise given that animal rights advocates have clearly stated that humans should not utilise animals for their own benefit (Rudy, 2011). Such a position sits at odds with the utilisation of animals in human leisure. The animal rights position may even question the volunteer work of people such as Sheila Scutter (Chapter 8) who spend large amounts of their free time and energy in the interests of freeing bears in Asia. This is because volunteers may be said to gain at the expense of the animals they are volunteering to help.

It is important to differentiate between the conceptual position that is 'animal rights' and the 'rights' of animals to welfare. The animal rights position is an inherently humanistic one that offers the opportunity for humans to impose their idealistic thoughts regarding animals, their needs, behaviours, and habitats onto animals. The active agency and sentience of animals may be overlooked, or placed in a position that is subservient to the views of humans. What the chapters in this book are focused on is the right of animals to welfare, including leisure, and the obligations of humans (a concept forwarded by Broom (2010), among others) as those in a position of power to ensure these welfare needs are met. This is arguably a biocentric position that places the animal at centre stage. It requires listening to animals and understanding their needs. More discussion on this is provided later in the chapter.

Animals at work and play?

Wilkinson, in her discussion of the life of Camilla the chameleon (Chapter 6), brings the issue of wild animals at work in human leisure to the fore, but the same point can be seen in virtually all of the other chapters. If we recognise animals as sentient beings then it becomes obvious that while they are in leisure attractions such as zoos, placed on display for humans, these animals must be 'at work' as they are not merely objects. It is important to recognise that such 'work' is not restricted to zoos, but can occur wherever humans seek wild animal experiences. The potential stresses of this work are identified in Chapter 2, by Ferguson and Litchfield. However, this chapter is set within a traditional construct of the impacts of human leisure on animals. Positioning animals as being 'at work' offers opportunities to reconceptualise the pressures on these animals, and to consider new and different approaches to responding to such stresses.

The need for recreation, the opportunity for recuperation and relaxation away from the stresses associated with work, is well documented in relation to humans (Bammel & Bussus-Bammel, 1996; Kelly, 1981). Could this research be beneficially applied to animal welfare? When raising this question care needs to be taken in applying the human construct of work to animals. This is primarily because humans can be seen to generally engage 'voluntarily' in work, whereas the same cannot be said of most, if not all, wild animals working in the human leisure experience. It is recognised that the pressures of needing to gain income in order to meet basic survival needs can be seen as significantly reducing the 'voluntariness' of human work (Beck, 2000). However, even where humans are under pressure to engage in labour there is still normally an element of negotiation and choice that can be perceived to distinguish the experiences of human and animal labour. Yet connecting the idea of animal labour with a concomitant understanding that human labourers require leisure indicates that animal workers have a need for the same. This leads to the need to recognise that recreation and leisure are not the same. Recreation is at root, concerned with re-creating the individual through recuperation. In this way, it is a break from work that enables a refreshment of the individual before he/she goes back to work. Recreation is therefore directly beneficial to the work environment. Leisure, in contrast, is not designed to reinvigorate the workforce but to allow for self-development. What this suggests is that leisure and recreation are different, though potentially related to the concept of 'enrichment' (as explored by Scutter and Young in Chapter 8 with regard to still captive bears in Chapter 8).

Leisure and freedom

Carr (2017) has discussed the position and importance of freedom to conceptualisations of leisure, recognising that while leisure includes issues of non-work time, and particular types of activities, it is inherently associated with both a *freedom from* and a *freedom to be*. It is these freedoms that enable the individual to find themselves and engage in personal development for the sake of the self. Such conceptualisations of leisure have all been made with only humans in mind. However, there is enough evidence regarding the play animals engage in (Burghardt, 2005; Peterson, 2013) to begin to question whether leisure is only a human need or desire. This is based on the recognition that play can be a part of leisure (Dumazedier, 1974; Kaplan, 1960).

Ideas about animals and play are raised in Chapter 8, by Scutter and Young, when looking at the environmental enhancements and enrichment provided to captive bears. But the work of Wilkinson in Chapter 6 (focused on Camilla the captive chameleon) and the ideas raised by Appleby and Smith (Chapter 11 – 'Do wild canids kill for fun?') and Yerbury and Boyd (Chapter 9 – the impacts of human leisure on dolphins) take this further to question whether animals as sentient beings need and desire leisure as a form of self-expression and development. This is a question that is loaded with potential anthropomorphic pitfalls, is not restricted to wild animals and is only just beginning to be explored. If animals do need and wish to have this type of leisure then the question of freedom for these animals is raised. Carr (2017), amongst others, has recognised that even humans do not have absolute freedom, being social creatures. However, in order to experience leisure that is inner-directed and targeted on the search for self, a form of freedom is a requirement.

This leads to the question of whether such leisure is achievable for wild animals in captivity where any activities or options are structured and controlled by humans. Humans, as in the case of Scutter and Young's volunteers, may feel they have given freedom to the animals but this sits in opposition to the notion that freedom can only ever be taken, not given. Stirner for example, states that 'the man who is set free is nothing but a freed man, a libertinus, a dog dragging a piece of chain with him: he is an unfree man in the garment of freedom' (in Leopold, 1995, 152). Consequently, Stirner noted, 'all freedom is essentially – self-liberation – that I can have only so much freedom as I procure for myself by my ownness' (ibid.). This view is shared by Nietzsche (1967, 412), who stated that 'Individuals liberate themselves'. While Nietzsche and Stirner were clearly talking about humans, it may be argued that the same can apply to animals. This raises questions about whether in providing wild animals in captivity with leisure experiences that are defined and bounded by humans we are preventing them from evolving, or self-developing. Are we guilty of stymying the development of individuals and entire species in our rush to situate animals in a timeless space that gives them the ability to play in human-defined acceptable ways but deprives them of the opportunities for leisure that can enable self-development and expression? While this discussion may be focused on captive wild animals, the same issues and concerns can be applied to their non-captive counterparts and the control humans' exercise over them when deciding what is 'proper' leisure for these animals.

By questioning the nature of the leisure experiences available to wild animals we must question what people are doing in the name of ensuring these animals have a high quality of life through the provision of leisure opportunities. Without wishing to negatively anthropomorphise (see Carr (2014) for a discussion of the potential positives of anthropomorphism), it may be suggested that there is a parallel here to discussions about the provision of play opportunities for children. This literature has noted and critiqued an increasing tendency by adults to control and manipulate the leisure opportunities children are able to engage in. Children's play has become defined by adults who have specific agendas for that play, albeit overtly positive ones incorporating ideas of the development of children into productive members of (adult) society (Carr, 2011). Less positively viewed, it can be argued that children's play is structured by adults in such a way as to prevent them from bothering adults. Either way, such play is not constructed by children for children. The result is that there is little freedom in children's play that is constructed and policed by adults. This limited play has been criticised for its inability to allow children to take control of their own lives and to develop and test their own cognitive, physical, and social abilities (Honore, 2004). In effect, such play is hindering children's potential to search for and develop a sense of their own self and instead imposes a way of being on them that is controlled by the powerful of society (that is, adult humans).

A parallel argument can be made that play opportunities for wild animals in spaces constructed and regulated by humans are similar to those for children in the contemporary society. Adults are generally well meaning in the provision of children's play opportunities and arguably, so are the human providers of play opportunities for animals. However, no matter how well meaning, human actions need to be examined critically, with an eye to determining whether options provided are what is best for animals from the perspective of animals themselves, or merely what is best for the animals as far as humans are concerned. This speaks to the need to examine animal welfare from the viewpoint of animals rather than from that of humans. Humans often have set ideas about appropriate animal behaviour but in this framing we may be robbing animals (individually and potentially collectively at times) of the opportunity to evolve; something that we hold dear as a right of all humans. This issue can be related to the debate about approaching wild animals versus being approached by wild animals. Should we continually seek to reinforce the message that wild animals and humans should be separated from one another, primarily for the benefit of the former, or should we consider the possibility that some wild animals may wish to explore interactions with humans for their own benefit. This issue is related to Yerbury and Boyd's work in Chapter 9.

Ethical hunting?

Another thorny question raised in this book, particularly in Chapters 10 (by Mordue and Wilson), 11 (by Appleby and Smith) and 12 (by Bray *et al.*) is related to hunting. Lovelock and Lovelock (2013) are among those who have argued that ethical hunting of animals is both possible and desirable. Conversely, Goodall (forthcoming), amongst others, has argued passionately against all forms of hunting. What is apparent is that as animals are recognised as sentient beings in increasing forums, more nuances and complexities related to animal hunting are emerging. The relationships between groups of animals, and the rights of each to freedom from, and freedom to, hunt needs to continue to be unpacked. The reality that animals hunt other animals needs to not be overlooked in these explorations; to do so would itself be a form of human ideological imposition on the animal kingdom.

It is possible to question whether linking the term 'ethical' to the activity of hunting is anything more than a public relations stunt intended to deflect criticism by the anti-hunting lobby. In theory, it may be possible to claim to engage in ethical hunting if by doing so we are saving animals from a life of intensive farming or protecting the natural environment by removing introduced species from it. However, such arguments are filled with problems. Such a position prioritises species conservation over individual animal welfare. Furthermore, it maintains the power of humans, deciding which animals are to live and which are to die, which have more value and which have less. Finally, the argument in favour of removal of introduced species only works if such leisure hunting actually does what it says instead of simply *harvesting* a number of introduced animals on an ongoing basis. This said, arguments in favour of ethical hunting, of removing the need to rely on factory farming and of helping to remove damaging introduced species from ecosystems can be persuasive. Until, that is, we recognise that we do not argue for humans – the most destructive of all animal species – to be removed in this way!

Appleby and Smith (Chapter 11) viewed the issue of leisure hunting from a different perspective: that of the animal as hunter. The animal as leisured hunter brings all sorts of questions about the rights of wild animals to hunt domestic and domesticated animals into play. This brings into sharp relief the power dynamics that colour all relations between humans and animals. It is these dynamics that need to be recognised when we examine leisured hunting, by wild animals and humans. The reality is that humans are the dominant species on the planet, the ones in the position of power to which all other animals are subordinate. It is within this uncomfortable reality that leisured hunting must be understood.

If ethical hunting by humans is to exist it must be based on the welfare of individual animals as well as species and arguably should not play God by prioritising one species over another. When the desires of humans are placed before the welfare of animals such hunting surely cannot be called ethical. Likewise, we must question whether it is ethical to prevent wild animals from engaging in leisured hunting, if animals wish to do so.

Human leisure needs with regard to captive animals

Following on from the point noted above (that animals exist, albeit to varying distances of relationship, in a human-dominated world) we would assert that it is not enough to simply recognise the welfare needs and rights of captive wild animals. The chapters by Curtin and Green (Chapter 4) and Carr (Chapter 5) attest to this argument. Enmeshed with animal welfare and wellbeing concerns are human leisure desires in relation to wild animals. As both of these chapters show, people have a strong and enduring desire to not only see wild animals but to get as close as possible to and even interact (albeit it in a one-sided, humancentric sense) with them. As Ferguson and Litchfield (Chapter 3) indicate, there is the potential for benefits to accrue from interactions between wild animals and humans but there are also potential problems associated with such practices. The issue facing leisure attractions is how to prevent the negative behaviours identified in Chapter 5 while not adversely impacting the potential benefits of animal–human interaction noted in Chapter 3. This is arguably the key challenge that leisure attractions like zoos and aquaria need to overcome in the face of ongoing calls for their closure on the basis of animal welfare concerns.

Those we love, in places we know

It has been good in editing this book to see work in the field of animals and leisure reaching beyond the traditional focus on the animals that we as humans tend to like the most. Instead, we have seen chapters focused on reptiles (specifically a chameleon), fish and dingoes. However, there is still much more work needed that explores and recognises the needs of the wild animals we do not love (including those we hate and those we simply ignore). The welfare of these animals is just as important, at both the individual and species level, as the welfare of those animals we love. Yet animals that are not widely loved are arguably more likely to see their welfare needs ignored or flatly rejected.

While the geographical spread of authors and research presented in this book is good, we are conscious that it does not include more work from outside of the developed and English-speaking nations of the world, with the chapter by Rizollo and Bradshaw the only one to focus attention on a non-western setting. The experiences of wild animals as sources of human leisure and their access to leisure in non-western, Anglo-Saxon settings need more analysis and discussion.

The breadth of disciplinary settings of the authors of this book demonstrates how integrated leisure studies can be across the research environment. This harkens back to comments made in Chapter 1 with regard to the emergent strength of leisure studies, not as a discipline but as a multi-disciplinary meeting ground. The diverse array of disciplines represented by the authors also speaks to the trans-disciplinarity of human–animal studies. The fostering of these multi-disciplinary perspectives on leisure studies and animal welfare needs to be encouraged and nurtured into the future as a means of addressing the complex questions raised by the intersections of human and animal leisure.

Listening to animals

To be able to prioritise the needs and position of animals there is clearly a need to *listen* to them. Only sentient, socially active beings, after all, are capable of communicating. We are not, of course, referring to speaking or listening here as they have been constructed by humans. Animals are not humans and to suggest otherwise is to engage in what we would see as misdirected anthropomorphism.

There is a need to engage with the 'other', in this case the non-human animal, and to listen to them not necessarily solely on their terms but at the meeting ground that exists between species. Such meeting grounds are arguably broader and better understood between species that have a long history of friendly cohabitation with humans. Indeed, Bekoff (2007) and Horowitz (2009) have argued that if we as humans are prepared to listen then it is entirely possible to understand what the domestic dog is trying to tell us. The meeting ground between humans and wild animals may be narrower and less clearly understood but this does not mean that we cannot, if we are willing to try, reach a level of cross-species understanding.

It is important to recognise the inequitable power dynamics discussed earlier in this chapter in our attempts to understand and talk with other species. This means realising that listening and talking are two-way processes but that one side may want to listen or talk more than the other. While this may be the case, it is important that when listening and talking to other species we constantly allow the 'other' to talk and listen to us so that it is possible to engage in conversation that is two-way communication. For more detailed discussions about listening and talking to 'other' species, readers are invited to look at the work by Young and Carr (forthcoming, b) and Carr (2014), amongst others.

Speaking of edited books

Both of the authors of this chapter exist in national academic settings where the value of edited books (and even authored monographs and textbooks) is being questioned as emphasis is placed on peer-reviewed journal articles as the primary publishing outlet of research. This is a situation we are sure that many colleagues around the world can identify with. This obsession, for obsession is what it can feel like, is being driven, at least partially, by national tertiary education funding mechanisms. As a result, everyone has heard the stories where universities *encourage* staff to publish only in specific journals (based on ranking, Impact Factor and/or field of study).

From a neo-liberal capitalist perspective, we can comprehend why universities feel the need to engage in this behaviour and equally understand why staff often feel compelled to 'play the game'. However, edited books offer the opportunity to bring together a diverse array of authors from disparate fields (health, volunteering, environmental studies, sociology, and tourism, to name a few, as evidenced by the authors of the chapters in this book) to produce a compilation of value beyond the sum of its parts.

This edited book, and its twin (Young & Carr, forthcoming, a), have been a joy to edit as the chapters have collectively and individually inspired, showing new ways of thinking and new areas of work to be explored. This is what edited books are able to do, and what journals are not capable of doing. As editor of the *Annals of Leisure Research*, Neil sees a wonderful array of articles that attest to the vibrancy of the leisure studies research field. Yet each article is just an individual snippet. An edited book offers the chance to move beyond singular voices to a polyphonic, choral approach. The contrasts, and indeed contradictions between authors, the diversity of stories, origins and perspectives combine to create a whole that is more than any one singular authorial voice. The result prompts new insights, suggesting alternative approaches and reveals surprising tangents and connections. In this way, edited books offer huge potential to forward a field of study and to have an impact beyond any one academic field. Consequently, we encourage authors, universities and funding bodies not to cast aside edited books.

Activism and research

There are doubtless other issues to be drawn from the chapters of this book than those summarised by us, the editors. One of the delights of an edited collection is that everyone may read the chapters, individually or collectively, and glean new and different insights from them. Consequently, the insights discussed in this chapter should be viewed only as some of many possibilities. Fundamentally, the sphere of animals (both wild and domestic) and leisure is one that is filled with a diverse range of research opportunities, virtually all of which have yet to be fully explored. Yet, we believe, such research is valueless if it does not grasp the concept that it should all be situated within the context of animal rights and welfare. The fundamental rationale for conducting research in this field should not be to simply allow academics to indulge themselves in research just for the pleasure of it, or in the pursuit of promotion. Nor should it be undertaken for the betterment of humans, collectively or individually. Rather, while all of these things may stem from research on animals and leisure the primary focus and underlying rationale must be animal welfare and rights.

References

Bammel, G. & Bussus-Bammel, L. (1996). *Leisure and Human Behavior* (3rd ed.). Brown & Benchmark Publishers. Madison, WI.

Beck, U. (2000). *The Brave New World of Work*. Polity Press. Malden, MA.

Bekoff, M. (2007). *Animals Matter*. Shambhala. Boston.

Broom, D. (2010). Cognitive ability and awareness in domestic animals and decisions about obligations to animals. *Applied Animal Behaviour Science* 126 (1): 1–11.

Burghardt, G. (2005). *The Genesis of Animal Play: Testing the Limits*. MIT Press. Cambridge, MA.

Carr, N. (2017). Re-thinking the relation between leisure and freedom. *Annals of Leisure Research* 20 (2): 137–151.

Carr, N. (2014). *Dogs in the Leisure Experience*. CABI. Wallingford, UK.

Carr, N. (2011). *Children's and Families' Holiday Experiences*. Routledge. London.

Dumazedier, J. (1974). Leisure and the social system. J. Murphy (ed.). *Concepts of Leisure: Philosophical Implications*. Prentice-Hall. Englewood Cliffs, NJ, pp. 129–144.

Goodall, J. (forthcoming). Sport hunting tourism. N. Carr & D. Broom (eds). *Animal Welfare and Tourism*. CABI. Wallingford, UK.

Honore, C. (2004). *In Praise of Slowness: How a Worldwide Movement is Challenging the Cult of Speed*. Harper San Francisco. New York.

Horowitz, A. (2009). *Inside of a Dog: What Dogs See, Smell, and Know*. Scribner. New York.

Kaplan, M. (1960). *Leisure in America: A Social Inquiry*. John Wiley & Sons. New York.

Kelly, G. (1981). *Leisure in Your Life: An Exploration*. Saunders College Publishing. Philadelphia.

Leopold, D. (1995). *Max Stirner. The Ego and Its Own*. Cambridge University Press. Cambridge.

Lovelock, B. & Lovelock, K. (2013). *The Ethics of Tourism: Critical and Applied Perspectives*. Routledge. London.

Nietzsche, F. (1967). *The Will to Power*. New York: Vintage Books.

Peterson, A. (2013). *Being Animal: Beasts and Boundaries in Nature Ethics*. Columbia University Press. New York.

Rudy, K. (2011). *Loving Animals: Toward a New Animal Advocacy*. University of Minnesota Press. Minneapolis.

Young, J. & Carr, N. (eds). (forthcoming, a). *Domestic Animals, Humans, and Leisure: Rights, Welfare, and Wellbeing*. Routledge. London.

Young, J. & Carr, N. (forthcoming, b). Domestic animals leisure, rights, wellbeing: nuancing 'domestic', asymmetries and into the future. J. Young & N. Carr (eds). *Domestic Animals, Humans, and Leisure: Rights, Welfare, and Wellbeing*. Routledge. London.

Index

Note: *italic* page numbers indicate tables, **bold** indicate figures.